Lecture Notes in Economics and Mathematical Systems

Managing Editors: M. Beckmann

Operations Research

103

D. E. Boyce A. Farhi
R. Weischedel

Optimal Subset Selection

Multiple Regression, Interdependence and Optimal Network Algorithms

Springer-Verlag
Berlin · Heidelberg · New York 1974

David E. Boyce, A. Farhi, R. Weischedel
Regional Science Department
University of Pennsylvania
Philadelphia, PA 19174/USA

Library of Congress Cataloging in Publication Data

Boyce, David E
Optimal subset selection: multiple regression.

(Lecture notes in economics and mathematical systems, 103)
Bibliography: p.
1. Electronic data processing--Regression analysis.
2. Electronic data processing--Mathematical optimization.
I. Farhi, A., 1944- joint author.
II. Weischedel, R., 1949- joint author.
III. Title. IV. Series.
QA278.2.B69 519.4 74-18097
ISBN 3-540-06957-7

AMS Subject Classifications (1970): 60H25, 62PXX, 90PAXX, 90B99, 90C50

ISBN 3-540-06957-7 Springer-Verlag Berlin · Heidelberg · New York
ISBN 0-387-06957-7 Springer-Verlag New York · Heidelberg · Berlin

Offsetprinting and bookbinding: Julius Beltz, Hemsbach/Bergstr.

Preface

In the course of one's research, the expediency of meeting contractual and other externally imposed deadlines too often seems to take priority over what may be more significant research findings in the longer run. Such is the case with this volume which, despite our best intentions, has been put aside time and again since 1971 in favor of what seemed to be more urgent matters. Despite this delay, to our knowledge the principal research results and documentation presented here have not been superseded by other publications. The background of this endeavor may be of some historical interest, especially to those who agree that research is not a straightforward, mechanistic process whose outcome or even direction is known in advance. In the process of this brief recounting, we would like to express our gratitude to those individuals and organizations who facilitated and supported our efforts.

We were introduced to the Beale, Kendall and Mann algorithm, the source of all our efforts, quite by chance. Professor Britton Harris suggested to me in April 1967 that I might like to attend a CEIR half-day seminar on optimal regression being given by Professor M.G. Kendall in Washington, D.C. I agreed that the topic seemed interesting and went along. Had it not been for Harris' suggestion and financial support, this work almost certainly would have never begun.

Even then, nothing happened for over a year; it was not until Ben Stevens and Bob Coughlin of the Regional Science Research Institute approached me in June 1968 with a problem of selecting the most important of some 100 metropolitan indicators that work was initiated. I suggested that interdependence analysis, not principal components analysis which they were considering, was the method of choice. They agreed to support an effort to develop the necessary programs; Andre Farhi set to work deciphering Beale, Kendall and Mann's article, and Ralph Weischedel expertly began to prepare the code. We are indebted to RSRI for this initial impetus and support. Regretably, Bob Coughlin's "best indicators" remain unselected, as events overtook us and we never prepared a program with sufficient capacity to accommodate his problem.

Later that summer, Dr. Richard Hickey approached Andre and Ralph with a problem he was having in interpreting a stepwise regression program. They suggested he try our newly coded optimal regression algorithm. In time, he produced some results which are highly suggestive of the merits of replacing stepwise regression with an optimal search algorithm; this experience is summarized in Chapter 2. The following spring (1969), when Resources for the Future agreed to support Dick Hickey's ongoing research, he enlisted the three of us to work on the project. During that second summer, we extended and refined the optimal regression and interdependence programs, and added a number of features to facilitate their use. The programs were carefully documented during this period; RSRI Discussion Paper No. 28 was issued in February 1969 followed by a users' guide in January 1970. The optimal regression program was used extensively during late 1969 and 1970 in our RFF research. During this period, Paul Slater ran some comparisons of optimal vs. stepwise regression which are presented in Chapter 2. We are pleased to acknowledge his contribution, and to thank RFF for its financial support.

During 1969, Andre pointed out that Beale's algorithm could be applied to the optimal network problem for which Allen Scott had developed a backtrack programming approach. While we might have been better advised to complete our unfinished statistical research, the challenge of building a better algorithm proved too much a temptation. The University's Urban Mass Transportation Administration research and training grant was tapped for a modest amount of funds for another summer's effort. We are grateful too for this additional upport.

The optimal network program fell into place quickly once we realized that simply replacing the size constraint with a linear budget constraint eliminated the initial paradox of how to compute optimal networks for combinations of network size (numbers of links) and budgets. What we failed to observe was the possibility that replacing one long link with two shorter ones might yield a better solution. Britton Harris helped us discover this flaw in our algorithm by developing a somewhat similar heuristic which found a better solution than one of our program's "optimal" networks. We would like to thank him for his encouragement and good advice (which we didn't always follow) throughout this period. The test problem which was the source of this discovery is one of four networks presented in Chapter 4. We are grateful to Maciej Luba for specifying these test problems and solving them.

A fourth summer (1971) was absorbed with completing the program writeups, chasing elusive program bugs, and generally completing the work. We were determined at this stage to complete the programs once and for all, and not be sidetracked into the pursuit of improvements in the basic algorithm. One such improvement that we did undertake was an alternative branching strategy for the network algorithm. It is not documented in this volume, although it is described in our Environment and Planning article, and the program (ONB) is available.

The programs and writeups were indeed completed in the summer of 1971. However, we decided the first priority was to publish our network algorithm which involved rerunning the four examples; to our dismay, this consumed our available time during 1971-72. The opportunity to undertake a visiting fellowship at the University of Leeds in 1972-73 presented what appeared to be fully ample time for me to assemble our algorithm notes, examples and program documentation into this monograph. Even so, pressure of other work prevented any concerted effort before June 1973. Several unfinished sections of the monograph were drafted and the complete work was edited and typed in June and July 1973. I am indeed grateful to the British Science Research Council for its support during this period. Andre and Ralph found time to proofread the final manuscript in August. The following twelve months were again incredibly absorbed by finding a publisher, preparing the flowcharts, completing the final copy and the other myriad details that always seem so incidental.

What seems clear in retrospect is the difficulty, as users of statistical methods and optimization techniques, of finding time to prepare a volume that we hope will be of considerable value to our research colleagues. Had we been mainly occupied as professional statisticians and programmers, we probably would have completed the work several years ago; however, we might have failed to recognize the importance of Beale's algorithm for substantive research. Therein lies the source of our determination to complete this work: our conviction, as yet not fully tested, that the optimal regression and interdependence algorithms should be widely applied in place of stepwise regression and principal components analysis when selection of subsets of variables is the objective.

In addition to the several individuals acknowledged above, we would like to thank E.M.L. Beale and J.D. Murchland for their incisive comments on our network algorithm. We alone remain responsible for errors, omissions and shortcomings. We are grateful to the University of Pennsylvania for making available computer time for writing and testing the programs. The U.S. Department of Transportation, through several contracts only partly related to this research, supported the typing of the volume. Diane Petch, a truly outstanding secretary, typed the original manuscript. Dorothy Yacek made the corrections, typed the flowcharts, managed other assorted details, and generally facilitated the work's completion. Godwin Odumah expertly drew the flowcharts.

Finally, I would like to acknowledge the contributions of my coauthors: to Andre, who specified the algorithms, developed the proofs, and explained them so clearly to us; and to Ralph who coded the programs without, it seemed, ever making an error, and wrote the documentation without ever remarking once how incredibly tedius a task it must have been.

Philadelphia, July, 1974 D.E. Boyce

<u>Table of Contents</u>

List of Tables

List of Tables

List of Figures

Program Availability

Individuals desiring to acquire the programs described in this monograph should send a 400-ft. (or longer) 9-track magnetic tape with standard label, specifying 800 or 1600 bpi, to D.E. Boyce, Regional Science Department, University of Pennsylvania, Philadelphia, PA 19174. Test decks corresponding to the examples are included. Subject to availability of computer time, no charge will be made.

1. INTRODUCTION

1.1 Orientation and Objective

The problem of selecting items (variables, transportation links, facility locations) from a larger set of all available or possible items, so as to optimize an objective function subject to a budget or size constraint, is a familiar problem in statistics, transportation research, operations research and related fields. Some cases of this general problem, such as the quadratic assignment problem, have been extensively studied. Others are virtually unknown.

The objective of this monograph is to make available the authors' computer programs and experience with algorithms for solving three optimal subset selection problems: (a) optimal regression analysis; (b) interdependence analysis; and (c) optimal network analysis. The first two problems are direct applications of the Beale, Kendall and Mann (1967) tree search algorithm. Solution of the third problem requires a generalization of their algorithm, replacing the constraint on the number of variables to be selected with a continuous budget constraint.

In the course of applying the Beale, Kendall and Mann algorithm to all three problems, descriptions of the algorithm were prepared that are much more detailed than other published versions. These descriptions should be useful both in understanding these algorithms and in constructing new algorithms similar to these. In addition, detailed user instructions and documentation for the three computer programs comprise about one-half of this volume. These should be very useful to users of the programs.

The importance of the optimal regression program for research must be strongly emphasized. During the past decade, stepwise regression analysis has been very widely applied, especially in the social sciences. Use of stepwise regression implies either an inductive research strategy or a desire for the convenience of a heuristic exploratory procedure. In any event, the subset of variables selected by stepwise regression may be substantially different, and inferior in terms of explained variance, to the best subset of variables. Few users of stepwise regression seem to appreciate the implications of this point. In relying on stepwise regression to select the best subset of variables, researchers may be seriously in error. If it is being used as a heuristic, stepwise regression is a highly limited procedure as it only examines one of many near optimal solutions that typically exist. For these reasons, widespread replacement of stepwise regression with Beale, Kendall and Mann's optimal regression algorithm is advocated both for finding best regression equations and for systematically exploring many good regression equations in a heuristic fashion.

1.2 Organization

The three algorithms and computer programs are presented according to a parallel format and structure in Chapters 2, 3 and 4. The basic outline of each chapter is as follows:

1. introduction
2. description of the algorithm
3. examples and case studies
4. strategy for using the computer program
5. program users' manual - description of input cards and output
6. programmer's manual - description of subroutines

Common terminology is used throughout to facilitate transfer of understanding and experience from one algorithm to another.

In Chapter 2, Optimal Regression Analysis, the general algorithm applied in all three cases is described in Section 2.2.1 followed by its application to multiple regression analysis in Section 2.2.2. A review of stepwise regression procedures is also included as Section 2.4 to clarify how they differ from optimal regression. Based on their experience with use of optimal regression, the authors offer fairly conclusive evidence of the importance of using this approach in Section 2.5. Strategies for using the program are introduced in Section 2.3 and described in detail in Section 2.6. Unusually detailed documentation of the input cards, output and subroutines are provided in Sections 2.7 through 2.11. Section 2.12 clarifies the differences among several definitions of the estimate of the square of the multiple correlation coefficient found in this and related stepwise programs.

Chapter 3, Interdependence Analysis, provides a parallel but somewhat briefer description of a procedure for selecting a subset of variables which best represents an entire variable set. The author's experience with this algorithm is quite limited, as reported in Section 3.3. Detailed user and programmer documentation are included as Sections 3.5 through 3.8.

Optimal Network Analysis, Chapter 4, represents the authors' principal innovation in this volume. Here, the basic algorithm is extended to a more general class of problems having a continuous budget constraint defined on the items being selected instead of simply a size constraint on the number of items in the subset. The application to a transportation network problem may be unfamiliar to statisticians. The case examined here is one of several ways of formulating the more general problem of how to identify the best transportation network. This version is not intended for application, except possibly in very simple situations. Instead, it represents an interim research result which will hopefully lead to improved procedures for addressing this issue; for related recent results on this problem, see Hoang (1973) and Steenbrink (1974a, 1974b).

Because of the size of realistic transportation networks, the performance of the algorithm as a heuristic procedure is an important result. It is for this reason that more detailed examples are provided in Section 4.3 than in corresponding sections of Chapters 2 and 3. However, this performance may not be sustained by further testing with weighted objective functions, larger networks, and budget costs not strictly proportional to distance. Therefore, one reason for making this program available is to permit other researchers to experiment with it.

It should be noted that this volume deals mainly with the authors' experience with three applications of Beale, Kendall and Mann's algorithm. Little attempt has been made to date to review the experience of other researchers with these procedures. Such reviews seem better published in journals in any event. Instead, the intent here is to provide a detailed statement of these algorithms and programs to facilitate their widespread application in research.

As noted in the Preface, the authors' "Optimal Network Problem: A Branch-and-Bound Algorithm" considers an alternative branching strategy which is not documented further here. This alternative was tested against the strategy described in Chapter 4, but was not competitive with it. Computer programs with identical input cards are available for both strategies. The first corresponding to the documentation in Chapter 4 is named ONA; the second is called ONB.

A detailed Table of Contents and Index are included to facilitate the use of this monograph as a reference and user's manual for the algorithms and associated computer programs.

2. OPTIMAL REGRESSION ANALYSIS

2.1 Introduction

Selection of variables for inclusion in a multiple regression equation is a familiar problem in empirical research. Desirably, theories or hypotheses provide guidance in specifying which variables to include. Often, however, several variables are available for measuring a particular attribute, and little basis exists for a clear-cut choice among them. Such situations arise frequently in the social sciences, and often also in engineering.

Stepwise regression analysis is a convenient and widely used technique in this situation. Stepwise regression procedures add variables to a regression equation so as to maximize the increment in variance explained with each variable added, i.e. at each step. Certain programs can also delete a variable if its contribution to explained variance falls below a significant level.

It is well-known, as illustrated below, that stepwise regression analysis does not always succeed in selecting the best subset of N variables from the set of available p variables, in the sense of maximizing explained variance. This result occurs because stepwise regression maximizes the increment to explained variance, and not explained variance itself. In some situations, the differences between the best subset of N variables and the stepwise solution for N variables may be substantial, not only in terms of differences in R^2, but perhaps more importantly in terms of the variables selected. In one example discussed below, the difference was crucial to the research findings. Situations may occur, then, in which the results obtained with stepwise regression procedures are very misleading.

In this chapter, an algorithm and computer program for selecting the best subset of N variables from a p-variable set is described. Sections 2.2 and 2.3 describe the algorithm and some basic features of the program. A brief review of related regression programs and a comparison with optimal regression are given in Sections 2.4 and 2.5. A detailed user's manual and program description complete the chapter.

2.2 Description of the algorithm

The optimal regression algorithm is a specific application of a quite general tree-search algorithm proposed by Beale, Kendall and Mann (1967); see also Beale (1970). This algorithm solves a general combinatorial problem of optimization on a discrete space. As will be shown, the principle of the algorithm is valid provided that the objective function of the problem only satisfies a rather weak condition. In return, however, the algorithm solves the problem in question efficiently only in certain conditions. These conditions are satisfied in the case of optimal regression analysis and problems of a similar nature, but may not hold for other problems that can be solved in principle by the algorithm.

For generality and clarity, it is worthwhile to describe the general algorithm, before applying it to optimal regression analysis, and to discuss under what conditions the algorithm may be efficient.

2.2.1 General Algorithm

Consider the following problem. A set R of p items $x_1 \ldots x_p$ is given. Let S_i, S_j.. be subsets of R, and let P(R) be the set of such subsets: $S_i, S_j \ldots \varepsilon P(R)$. A function f is defined on P(R) and takes its values in the set of real numbers, or any completely ordered set for that matter. Let $P_N(R)$ where $(1 \leq N \leq p)$ be the set of subsets of R with exactly N distinct elements. The problem is to maximize the function f on $P_N(R)$. If the elements of $P_N(R)$ are denoted S_i^N, S_j^N.. this amounts to finding an element of $P_N(R)$, say S_*^N, such that $f(S_*^N) \geq (S_i^N)$ for any $S_i^N \varepsilon P_N(R)$.

The function f is assumed to satisfy the following condition:

$$f(S_i) \geq f(S_j), \quad \text{if } S_i \supseteq S_j. \qquad \text{(Condition 1)}$$

As noted before, this is a rather weak condition. The validity of the algorithm depends only on Condition 1.

The algorithm consists of an exhaustive search in $P_N(R)$, constrained by bounding rules and guided by a search-ordering procedure. The bounding rules allow certain sets to be discarded without the value of their objective function being computed. The search-ordering procedure is used to choose the next set to be examined, among all those that cannot be discarded by the bounding rules. The efficiency of the algorithm depends crucially on the efficiency of the bounding rules in discarding a large number of sets; moreover, the efficiency of the bounding rules depends to some extent on the search-ordering procedure. However, choice of the search-ordering procedure is otherwise independent of the bounding rules and, to this extent, inessential. For these reasons, the nature of bounding rules is described first.

Bounding procedure The unconditional threshold of an item x_k is defined as

$$U(x_k) = f(R - \{x_k\}).$$

As a consequence of Condition 1, $U(x_k)$ is the highest value the objective function can attain for all subsets that do not include item x_k. Suppose that the best known value of the objective function on $P_N(R)$ is for S_c^N. Then, if $f(S_c^N) \geq U(x_k)$, the value of the objective function can be improved only if the item x_k is included. Hence, sets that do not include x_k can be discarded.

The conditional threshold of an item x_k given the items $x_{k_1}\ x_{k_2} \ldots x_{k_\alpha}$ is defined as

$$C(x_k \mid x_{k_1} \ldots x_{k_\alpha}) = f(R - \{x_k, x_{k_1} \ldots x_{k_\alpha}\})$$

Obviously

$$\begin{aligned} C(x_k \mid x_{k_1} \ldots x_{k_\alpha}) &= C(x_{k_1} \mid x_k, x_{k_2} \ldots x_{k_\alpha}) \\ &= C(x_{k_2} \mid x_k, x_{k_1}, x_{k_3} \ldots x_{k_\alpha}) \\ &\ \vdots \end{aligned}$$

Suppose that the best known value of the objective function on $P_N(R)$ is for S_c^N. Suppose that the sets excluding $x_{k_1} \ldots x_{k_\alpha}$ are now to be examined, other sets including them having been examined previously and rejected, for instance. Then, if

$$f(S_c^N) \geq C(x_k \mid x_{k_1} \ldots x_{k_\alpha})$$

the value of the objective function on sets excluding $x_{k_1} \ldots x_{k_\alpha}$ can be improved only if the item x_k is included. Hence the sets excluding $x_k, x_{k_1} \ldots x_{k_\alpha}$ can be discarded. This, again, follows only from Condition 1.

Thus, the bounding rules use the threshold concepts in order to discard various subsets. At each step of the search, the next step to be taken is restricted in this manner. However, this next step is not completely specified by the bounding rules, but also by the search-ordering procedure. The latter can be chosen arbitrarily provided only that it is exhaustive in principle; in other words, it must order the search among all the subsets that cannot be discarded on the basis of the bounding rules. The algorithm stops when every subset that has not been examined can be discarded by the bounding rules. In this fashion, the optimality of the result is guaranteed.

Search-ordering procedure The efficiency of the bounding rules clearly depends on the best known value of the objective function. Indeed, the higher the value of $f(S_c^N)$, the more subsets are likely to be discarded by using the bounding rules. Since the best known value $f(S_c^N)$ increases as the search proceeds, the efficiency of the bounding rules is enhanced during the search. If the search-ordering procedure is chosen carefully, the efficiency of the bounding rules can be further enhanced and the algorithm made to operate faster, provided that the search-ordering procedure itself is not too cumbersome. The choice of an adequate search-ordering procedure depends therefore on the structure of the specific problem to this extent.

The search-ordering procedure chosen by Beale et al. for the optimal regression algorithm seems to be well-suited for a large class of problems; its main lines can be approximately described as follows:

1. Begin with an arbitrary subset, say $S_o^N = \{x_{\alpha_1} \dots x_{\alpha_N}\}$;
2. Delete the item that causes the smallest decrease in the value of the objective function, and that has not yet been deleted;
3. Add the item that causes the largest increase in the value of the objective function. If the result is larger than $f(S_o^N)$, retain the new set as the best known set, and proceed as with S_o^N; i.e. return to step 2. If not, carry on the procedure of deleting one item at a time from S_o^N. If all the items have been tried without success, start dropping two items at a time from S_o^N, etc.

The search-ordering procedure and the bounding rules are completely described by the flowchart, Figure 2.1, in which:

Figure 2.1 - Flowchart for the General Algorithm

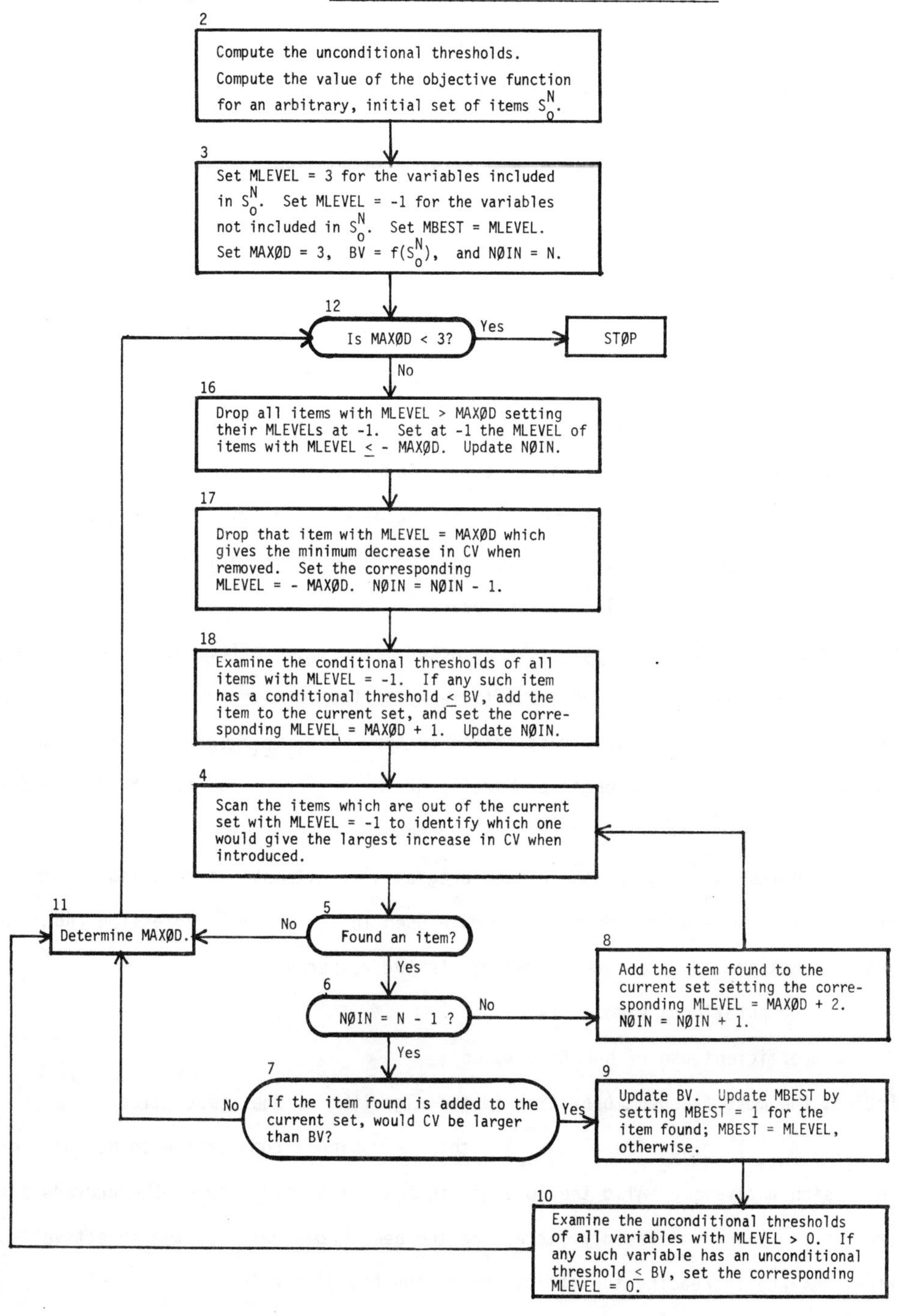

1. MBEST is an array of integers, one associated with each item denoting the best set found so far;
2. MLEVEL is an array of integers, one associated with each item recording the subset under examination and the ones that have been examined;
3. BV is the best known value of the objective function;
4. CV is the current value of the objective function for the set under examination;
5. MAXOD is the maximum odd value in the MLEVEL array;
6. NOIN is the number of items with MLEVEL nonnegative.

If stages 10 and 18 of the flowchart are eliminated, only the search-ordering procedure is represented; one can check that an exhaustive search would then be carried out. Finally, the search-ordering procedure can be characterized as a local optimum seeking procedure, in view of the criteria used in stages 17 and 4. (These criteria are very similar to the ones used in the backward elimination program and the forward selection program of Efroymson, 1962.) The starting point of the algorithm may also substantially influence its speed.

<u>Efficiency of the General Algorithm</u> As noted before, the efficiency of the general algorithm depends crucially on the efficiency of the bounding rules. These rules are valid under a rather weak restriction on the objective function, Condition 1. One may be able to use stronger properties of the objective function to devise more efficient bounding rules. This clearly depends on the problem at hand. In the optimal regression program and related programs, no attempt has been made in this direction. The search-ordering procedure used in the optimal regression and related programs has proved to be quite adequate in two respects:

1. computational simplicity
2. efficient use of bounding rules

Indeed, the most frequent operation in the algorithm is the computation of $f(x_{\alpha_1} \ldots x_{\alpha_{k+1}})$, given $f(x_{\alpha_1} \ldots x_{\alpha_k})$; this is a simple computation to perform in regression analysis. Also the local optimum seeking character of the procedure increases the chance of rapidly increasing the best known value of the objective function, thereby increasing the efficiency of the bounding rules.

2.2.2 Optimal Regression Algorithm

In optimal regression analysis, the "items" are the exogenous variables, and the objective function is the sample estimate (R^2) of the square of the multiple correlation coefficient (ρ^2) of the endogenous variable on the set of exogenous variables under consideration. Clearly, if the estimate of the multiple correlation coefficient is defined as the unadjusted maximum likelihood estimate, the objective function satisfies Condition 1. The general algorithm can therefore be used and guarantees optimal results: for a given N ($1 \leq N \leq p$), where p is the total number of exogenous variables, the subset of N exogenous variables yielding the highest R^2 with the endogenous variable is found.

In the case of optimal regression analysis, no simple criterion is available for comparing two regression equations including N and N' exogenous variables for $N \neq N'$. Therefore, one cannot be expected to decide a priori on an adequate value of N. For this reason, the optimal regression algorithm is programmed to enable the user to obtain the optimal regression equation for any increasing sequence of values of N. Figure 2.2 shows the flowchart of the algorithm for computing the optimal regression equations for specified values of N.

The most frequent operation in the algorithm is the computation of changes in the objective function when variables are added or deleted from the set under consideration. This computation can be carried out by pivot operations. Let

$$X = \begin{pmatrix} x_{11} \cdot \cdot & x_{1p} & x_{1,p+1} \\ \cdot & & \\ \cdot & & \\ \cdot & & \\ x_{n1} \cdot \cdot & x_{np} & x_{n,p+1} \end{pmatrix}$$

be a data matrix of n observations on p+1 variates; let $A = (a_{jk})$ be the sample estimate of the (p+1) x (p+1) coveriance matrix and $R = (r_{jk})$ be the sample estimate of the (p+1) x (p+1) correlation matrix. The following operations are now defined on the correlation matrix:

1. The transformation P_q on the matrix R (q = 1...p) is called "pivoting in the variable q": $R' = P_q(R)$

Figure 2.2 - Flowchart for Optimal Regression Analysis

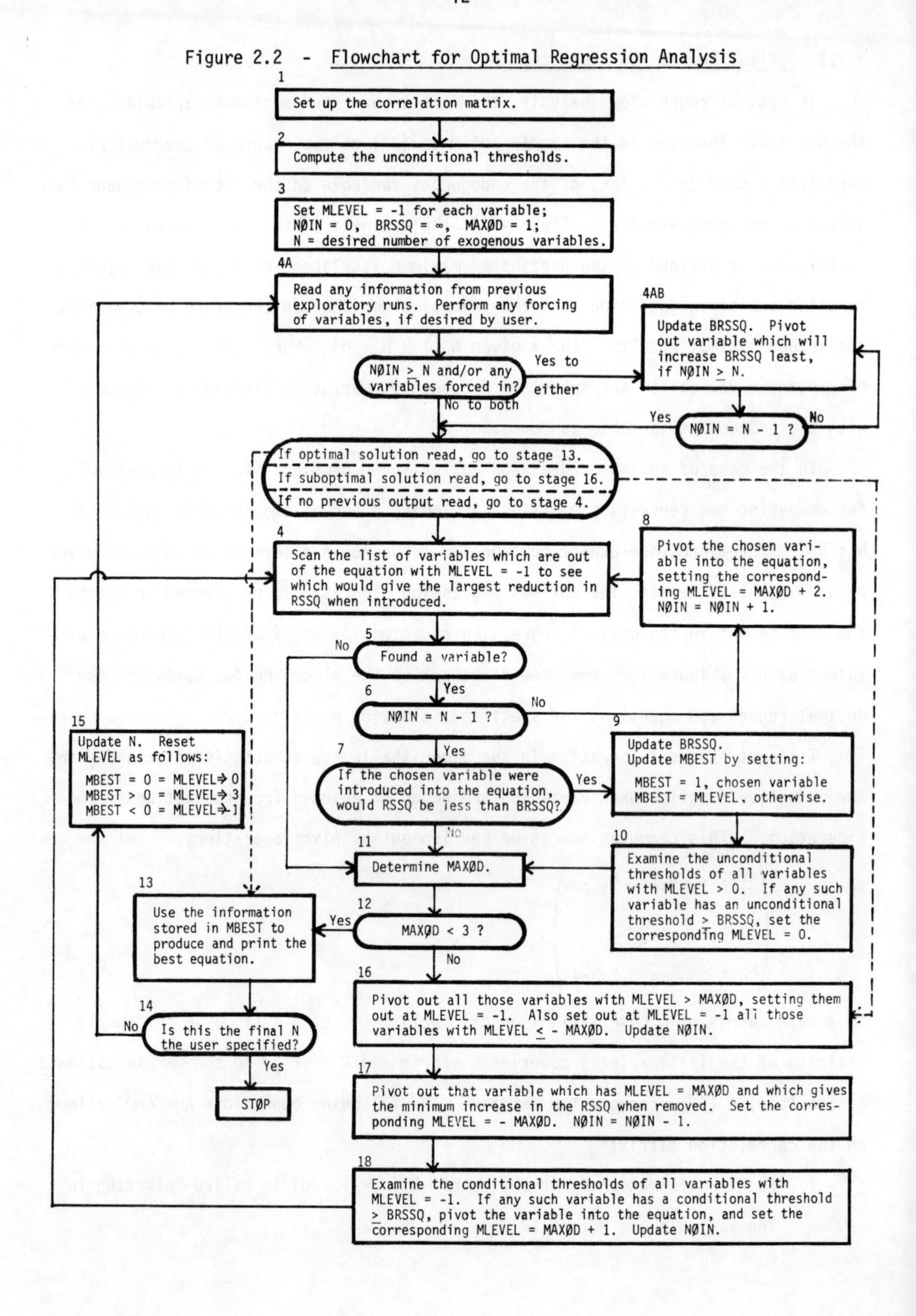

$$r'_{qq} = -1/r_{qq}$$

$$r'_{qj} = r'_{jq} = -r_{qj}/r_{qq}$$

$$r'_{jk} = r'_{kj} = r_{jk} - \frac{r_{jq}r_{qk}}{r_{qq}}$$

This transformation is not defined when $r_{qq} = 0$.

2. The transformation Q_q on the matrix R (q = 1...p) is called "pivoting out the variable q": $R' = Q_q(R)$

$$r'_{qq} = -1/r_{qq}$$

$$r'_{qj} = r'_{jq} = r_{qj}/r_{qq}$$

$$r'_{jk} = r'_{kj} = r_{jk} - \frac{r_{jq}r_{qk}}{r_{qq}}$$

The above transformation is not defined when $r_{qq} = 0$.

The successive application of the two transformations P_q and Q_q on R results in the original matrix R: $Q_q(P_q(R)) = R$.

To obtain the regression equation of x_{p+1} on say $x_{\alpha 1} \ldots x_{\alpha N}$, it is necessary to pivot in successively the variables $x_{\alpha 1} \ldots x_{\alpha N}$. Let R′ be the resulting matrix: $R' = P_{\alpha N} \ldots P_{\alpha 1}(R)$. To avoid the difficulties of near collinearities, in this pivoting process if a pivot element r_{qq} is smaller than a given level of tolerance ε, the corresponding variable is not pivoted in.

The following standard definitions are also needed to understand the algorithm:

1. Estimate square of the multiple correlation coefficient: $R^2 = 1 - r'_{p+1,p+1}$

2. Estimate of standard error on R : $\frac{(4R^2(1-R^2)(n-N-1)^2)^{\frac{1}{2}}}{((n^2-1)(n+3))^{\frac{1}{2}}}$

3. Estimate of regression coefficients of x_{p+1} on $x_{\alpha 1} \ldots x_{\alpha N}$:

$$\beta_{\alpha 1} = -r'_{1,p+1} \frac{(a_{p+1,p+1})^{\frac{1}{2}}}{(a_{\alpha 1,\alpha 1})^{\frac{1}{2}}}$$

$$\vdots$$

$$\beta_{\alpha N} = -r'_{N,p+1} \frac{(a_{p+1,p+1})^{\frac{1}{2}}}{(a_{\alpha N,\alpha N})^{\frac{1}{2}}}$$

$$\beta_0 = \bar{x}_{p+1} - \beta_{\alpha 1}\bar{x}_{\alpha 1} \ldots 1 - \beta_{\alpha N}\bar{x}_{\alpha N}$$

where $\bar{x}_i$ is the mean of the variable x_i.

4. Estimate of standard error on β_{α_k} $(k = 1 \ldots N)$: $\frac{a_{p+1,p+1}}{a_{\alpha k,\alpha k}} \cdot \frac{1 - R^2}{1 - S^2_{\alpha_k}} \cdot \frac{1}{n-N-3}$

   where $(1 - S^2_{\alpha k})$ is the (α_k, α_k) element of the matrix derived from R' by pivoting out the variable x_{α_k}.

5. F-ratio is equal to: $F = \frac{n-N-1}{N} \cdot \frac{R^2}{1 - R^2}$

6. Standard error of the estimate: $E = (\frac{1 - R^2}{n-1} a_{p+1,p+1})^{\frac{1}{2}}$

7. Estimate of residual sum of squares: $RSSQ = (1 - R^2)a_{p+1,p+1}$

8. The best residual sum of squares, BRSSQ, is the smallest value of RSSQ over all possible subsets of N variables.

For a detailed description of the optimal regression flowchart, see Section 3.2.2. The interdependence algorithm described there is exactly the same as the optimal regression algorithm, except for the definitions of RSSQ, BRSSQ and the pivoting procedure which are explained above.

2.3 Strategies in Using the Program

A summary of the strategic questions the user faces is included here for the general reader. In addition, this section serves as an introduction to the more detailed discussion of these questions presented in Section 2.6.

2.3.1 Alternate Ways of Stopping the Program

Experience with the program shows when N increases beyond a certain value, some regression coefficients become insignificant (according to a simple Student-Fisher t-test); eventually, the F-ratio decreases below a significant level. One is not usually interested in such equations. Therefore, it may be desirable to be able to stop the program when t or F become insignificant. However, it is not certain that higher values of N are also insignificant; therefore, one can pursue the computation for a few additional values of N and have the program stop only if these results persist. The program allows maximum flexibility with respect to these options. In addition, the program includes an option for computing and printing the residuals and for stopping the program if the Durbin-Watson test shows the existence of autocorrelation.

Since the F-ratio is equal to $\frac{n-N-1}{N} \cdot \frac{R^2}{1-R^2}$, the regression equation that maximizes R^2 for a given value of N also maximizes F for this value of N. Therefore, the program can be used to find the equation with the maximum level of significance of F for all values of N. If the level of significance of F-ratio is accepted as the criterion of comparison of two equations, whatever numbers of exogenous variables they contain, the program allows one to identify this equation. More generally, the equation with N variables obtained by the optimal regression algorithm is also optimal for any objective function of R^2 and N, say $h(R^2, N)$ such that $\frac{\partial h}{\partial R^2}(R^2, N) > 0$. In particular, the equation obtained maximizes the usual "corrected" estimate of the square of the multiple correlation coefficient; see Section 2.12.

2.3.2 Options for Using the Program

Although the optimal regression algorithm is much less time consuming than an exhaustive search, it can be quite cumbersome if used in the form presented in Figure 2.2. Three notions may be applied in order to improve the program's efficiency:

1. Stop the computation if the level of significance of F falls below a level prescribed in the program input, or if one or more regression coefficients is insignificant.

2. Operate the program on a loose grid of values of N. For instance, if n = 20, use the program initially for N = 1, 6, 11, 16 and decide on the basis of the output (in particular the behavior of the F-ratio and the regression coefficients) that the interesting possibilities occur for, say, $6 \leq N \leq 11$; then use the program a second time for N = 7, 8, 9, 10. (The program is conceived in such a manner that only very short calculations are duplicated in restarting the computation.) For instance, if one is interested in the second run in regressions for N = 7, 8, 9, 10 and 12, then punched output from the first run permits the computer to read the results for N = 6 and 11, thus providing good starting points for N = 7 and 12.

3. Limit the number of iterations of the algorithm through stage 12. For each N one can limit the number of iterations the algorithm will perform. The best result obtained during this search is printed out with an indication whether the result obtained is guaranteed optimal or not. In addition, the program punches cards for use as inputs in a subsequent run permitting the computer to continue its computations exactly at the point where they were stopped for each N. Accordingly, the computer time lost in stopping the program for a given N after a limited number of iterations, looking at the output and deciding that more iterations for this value of N would be worthwhile, is minimal.

These three approaches to reducing the computer time used by the optimal regression algorithm can be combined in various ways as described in detail in Section 2.6. The program also allows one to constrain the search to equations including and/or excluding certain variables. The variables to be "forced in" or "forced out" of the equation can be different for different values of N.

2.4 Related Multiple Regression Programs

There are $\binom{p}{N}$ multiple correlation coefficients between a dependent variable and subsets of N variables chosen among p variables. If the equation yielding the maximum square of the multiple correlation coefficient were to be found for each N, an exhaustive search would examine 2^p possible equations. As seen above, the optimal regression algorithm discards only the possibilities that can be rejected on the basis of the bounding rules. Hence the optimality of the results is guaranteed. Although the algorithm is much less time consuming than an exhaustive search, if used without the refinements of Section 2.3, it is still more cumbersome than three other approaches commonly used for regression analysis. However, these approaches do not guarantee optimal results: forward selection programs; backward elimination programs; and stepwise programs; see Efroymson (1962) and Draper and Smith (1966).

The forward selection approach first selects the variable with the maximum correlation coefficient with the dependent variable; then, the variable which, given the first, most increases the multiple correlation coefficient is pivoted in, etc. The first step of this algorithm coincides with the first step of the optimal regression algorithm, but the results can diverge as soon as the second step is taken. The forward selection algorithm is much faster than the optimal regression algorithm but does not guarantee optimal results. Experience with the two programs shows that the results of the forward selection algorithm can be substantially improved by optimal regression analysis, but the time spent on improving the results may be substantial.

The forward selection algorithm is, in fact, a special case of the optimal regression algorithm. It can be obtained by limiting the number of iterations to one for each N. In this way stages 16, 17 and 18 are never used. Hence, by increasing slightly the number of iterations, one can achieve improvements in the forward selection results without spending too much time. In other words, the optimal regression algorithm allows maximum flexibility in the tradeoff between time and higher multiple correlation coefficients, or more generally the quality of the results.

The backward elimination algorithm starts with the equation including all the exogenous variables and drops the one that results in the least decrease in the objective function; then, given the remaining variables, the one that results in the least decrease in the objective function is dropped, etc. (see stage 17 of the optimal regression algorithm flowchart). The computation time required for this program is quite similar to the forward selection algorithm. For this reason, it seems preferable to use the optimal regression algorithm and limit the number of iterations, as greater flexibility in the tradeoff between time and quality is a decisive advantage.

The stepwise regression analysis operates on the same basic principle as the forward selection algorithm, but at any stage variables can be dropped out of the equation and replaced by others. The criterion used to drop variables is a partial F-ratio criterion, and its tightness can be set by the user. Thus, the range of possible equations explored by this program can be widened by the user; this also results in a rather flexible tradeoff between time and quality of the results.

The advantage of the optimal regression algorithm is that provided enough time, it does guarantee optimal results. This is not the case with the stepwise regression algorithm, unless the partial F criterion is adjusted in such a way that a truly exhaustive search is completed. To do so would use much more time than the optimal regression algorithm uses to achieve the same result.

In conclusion, the stepwise algorithm and the optimal regression algorithm can produce results of similar quality, using comparable amounts of time if one seeks to conserve computer time. If the best results are preferred, optimal regression is decisively preferable.

2.5 Case Studies of Optimal vs. Stepwise Regression

Perhaps the most convincing evidence for use of the optimal regression algorithm in place of stepwise regression and related procedures is actual research experience. The following examples are drawn from studies reported by Hickey et al. (1970a, 1970b, 1971); as these examples are based on only one data set, caution must be exercised in assessing the generality of the results.

A brief description of the research problem and the data may help in understanding the examples presented below. The research objective was to examine the statistical relationship, if any, between each of several specific chronic disease mortality rates for U.S. metropolitan areas and the concentration of several environmental chemicals in the metropolitan centers. Data were assembled for 38 metropolitan areas on 16 environmental chemicals, mainly atmospheric, and a large variety of mortality rates, only a few of which are considered here.

Although biological theory did suggest some hypotheses for testing, such as the relationship between lung cancer mortality and SO_2 and NO_2, initially a more exploratory approach was taken. A well-known stepwise regression program by Cooley and Lohnes (1962) was used. In one case (simplified slightly for purposes of this exposition), the following equation was the best stepwise regression in the approximate sense that, based on t-tests, each of the regression coefficients was significant; that is, all equations with three or more variables had one or more regression coefficients that failed to be significant at the $\alpha = 0.05$ level.

$$Y = 13.206 + \underset{(t=3.74)}{0.896\,X_1} - \underset{(t=2.51)}{1.821\,X_2}$$

This equation has $R^2 = 0.325$ with $F_{2,35} = 8.45$, which is significant at the $\alpha = 0.01$ level. However, theory suggested two different variables, X_3 and X_4, should be significant predictors. In fact, these two variables were entered by the stepwise program in the third and fourth steps, but the regression coefficient for X_1 was then no longer significant at the 0.05 level. To test the hypothesis suggested by theory, these two variables were entered into the equation first, followed by other variables in a stepwise fashion. The following very interesting result was obtained for three variables:

$$Y = -22.809 + \underset{(t=4.63)}{4.992\,X_3} + \underset{(t=4.90)}{2.038\,X_4} - \underset{(t=3.88)}{1.286\,X_5}$$

which has $R^2 = 0.553$ with $F_{3,34} = 14.01$, also significant at the 0.01 level. Not only was X_5, rather than X_1 or X_2, added by the program, but also R^2 was substantially higher, even compared to the three variable equation with X_1 and X_2. Therefore, it seemed choice of X_1 and X_2 for the equation led to suboptimal results, and more importantly, results that were less interesting from the viewpoint of the theory. The optimal regression algorithm later confirmed that the above equations for N = 2 and 3 were optimal, and that X_1 was also excluded from the optimal equations for N = 4 and 5.

Subsequently, a more systematic comparison of optimal and stepwise regression was performed using the stepwise program BMD 02R by Dixon (1968, p. 233). The results shown in Tables 2.1 to 2.4 illustrate some of the most interesting examples of differences between optimal and stepwise regressions; the correlation matrix for these regressions is shown in Table 2.5. These examples demonstrate two important points:

1. Stepwise regression does not always find the best variable subset for a given value of N;
2. Variables included in lower order optimal regressions are sometimes dropped in higher order optimal regressions.

Altogether, regressions for 17 dependent variables were computed; the overall results are shown in Table 2.6. There are 68 nontrivial comparisons shown for optimal vs. stepwise equations with 2, 3, 4 and 5 variables. In 24 of 68 cases, or over 35 percent, the optimal equation was different from the stepwise result. Whether this proportion could have been lowered by more careful choice of the parameter controlling the partial F-test in the stepwise program was not investigated. Considering the extent of the differences found, however, it appears that optimal regression should always be used in preference to stepwise regression in cases in which the best equations are desired.

2.6 Suggestions for a Strategy for Using the Program

With the type of data encountered in the social sciences, the number of variates, p, is often close to the number of observations, n. In this case, the regression of a dependent variable on a number of independent variables, N, close to p, will often be found to be insignificant, that is yielding a relatively small value of the F-ratio. The optimal regression program calculates regressions for a dependent variable such that the residual sum of squares is a minimum. By using bounding

Table 2.1

Optimal and Stepwise Regressions for Breast Cancer

Number of Variables	R^2	Regression Coefficients with Standard Errors					
		Vanadium	Titanium	Nitrogen Dioxide	Sulphur Dioxide	Cadmium	Copper
1-optimal	0.2946	1.028 (0.272)					
1-stepwise	0.2946	1.028 (0.265)					
2-optimal	0.4220	1.044 (0.250)	-2.046 (0.759)				
2-stepwise	0.4220	1.044 (0.243)	-2.046 (0.737)				
3-optimal	0.5768			5.658 (1.203)	2.000 (0.463)	-1.319 (0.369)	
3-stepwise	0.5275	0.750 (0.247)	-2.039 (0.676)	3.457 (1.254)			
4-optimal	0.6252			5.520 (1.150)	2.018 (0.443)	-0.992 (0.389)	-1.500 (0.754)
4-stepwise	0.5822	0.480 (0.269)	-2.099 (0.645)	3.429 (1.197)	0.983 (0.472)		
5-optimal	0.6525		-1.077 (0.701)	5.256) (1.141)	1.915 (0.439)	-0.785 (0.404)	-1.260 (0.756)
5-stepwise	0.6385	0.333 (0.263)	-1.429 (0.681)	4.615 (1.250)	1.535 (0.510)	-0.873 (0.391)	

Table 2.2

Optimal and Stepwise Regressions for Liver Cancer

Number of Variables	R^2	Regression Coefficients with Standard Errors					
		Sulphur Dioxide	Water Hardness	Nitrogen Dioxide	Chromium	Titanium	Cadmium
1-optimal	0.1524	0.366 (0.148)					
1-stepwise	0.1524	0.366 (0.144)					
2-optimal	0.2823	0.417 (0.140)	0.366 (0.149)				
2-stepwise	0.2823	0.417 (0.136)	0.366 (0.145)				
3-optimal	0.3650		0.398 (0.144)		0.847 (0.229)	-0.639 (0.256)	
3-stepwise	0.3501	0.473 (0.134)	0.338 (0.141)	-0.679 (0.360)			
4-optimal	0.4547		0.370 (0.136)	-0.788 (0.349)	0.976 (0.223)	-0.710 (0.243)	
4-stepwise	0.4310	0.369 (0.136)	0.344 (0.134)	-0.796 (0.346)	0.422 (0.195)		
5-optimal	0.5525		0.353 (0.125)	-1.117 (0.346)	0.847 (0.211)	-0.851 (0.230)	0.272 (0.106)
5-stepwise	0.5403	0.308 (0.126)	0.394 (0.123)	-0.887 (0.318)	0.772 (0.218)	-0.615 (0.223)	

Table 2.3

Optimal and Stepwise Regressions for Lung Cancer

Number of Variables	R^2	Regression Coefficients with Standard Errors: Particulate Sulphate	Arsenic	Nitrogen Dioxide	Lead	Manganese	Vanadium	Titanium
1-optimal	0.1740	2.759 (1.031)						
1-stepwise	0.1740	2.759 (1.002)						
2-optimal	0.3802	3.992 (0.979)	-0.932 (0.281)					
2-stepwise	0.3802	3.992 (0.951)	-0.932 (0.273)					
3-optimal	0.4653	3.156 (0.995)	-1.080 (0.273)	2.972 (1.317)				
3-stepwise	0.4653	3.156 (0.966)	-1.080 (0.265)	2.972 (1.278)				
4-optimal	0.5612		-1.021 (0.267)			2.504 (0.560)	0.969 (0.215)	-2.796 (0.764)
4-stepwise	0.5607	3.315 (0.890)	-1.100 (0.244)	4.137 (1.253)	-2.058 (0.768)			
5-optimal	0.6607		-1.274 (0.253)	3.305 (1.114)		2.507 (0.500)	0.717 (0.210)	-2.582 (0.687)
5-stepwise	0.6086	2.633 (0.920)	-1.342 (0.264)	4.900 (1.262)	-2.191 (0.739)	0.943 (0.476)		

Table 2.4

Optimal and Stepwise Regressions for Arteriosclerotic Heart Disease

Number of Variables	R^2	Particulate Sulphate	Titanium	Arsenic	Copper	Sulphur Dioxide	Zinc	Lead
		Regression Coefficients with Standard Errors						
1-optimal	0.2638	45.143 (12.569)						
1-stepwise	0.2638	45.143 (12.569)						
2-optimal	0.4286	52.163 (11.446)	-28.827 (9.072)					
2-stepwise	0.4286	52.163 (11.446)	-28.827 (9.072)					
3-optimal	0.4955	65.921 (12.376)			-24.061 (9.383)		-16.337 (6.465)	
3-stepwise	0.4953	60.623 (11.620)	-24.010 (8.944)	-7.288 (3.438)				
4-optimal	0.6079	47.237 (12.660)			-26.513 (8.431)	18.948 (6.230)	-20.512 (5.943)	
4-stepwise	0.5334	62.111 (11.378)	-19.859 (9.089)	-6.079 (3.435)	-15.891 (9.689)			
5-optimal	0.6244	46.017 (12.643)			-24.762 (8.528)	20.253 (6.303)	-17.964 (6.318)	-11.867 (10.349)
5-stepwise	0.5800	47.739 (13.335)	-17.679 (8.883)	-6.037 (3.309)	-18.784 (9.460)	11.947 (6.341)		

Table 2.5

Correlation Coefficients for Environmental Chemicals and Mortality Rates

Variables	Cd	Cr	Cu	Fe	Pb	Mn	Ni	Sn	Ti	V	Zn	NO_2	SO_2	$SO_4^=$	WH	As	Breast Cancer	Liver Cancer	Lung Cancer	Heart Disease
1. Cadmium (Cd)	1.00	0.46	0.47	0.35	0.44	0.41	0.30	0.54	0.38	0.18	0.68	0.39	0.46	0.36	0.00	0.50	-0.00	0.28	-0.04	-0.08
2. Chromium (Cr)		1.00	0.21	0.46	0.44	0.43	0.41	0.41	0.54	0.36	0.48	0.23	0.39	0.50	-0.09	0.17	0.09	0.36	0.24	0.07
3. Copper (Cu)			1.00	-0.02	0.27	0.26	0.10	0.17	0.36	-0.08	0.23	0.14	0.23	0.22	0.33	0.33	-0.22	0.09	-0.29	-0.25
4. Iron (Fe)				1.00	0.31	0.30	0.00	0.22	0.34	0.03	0.35	0.12	0.19	0.14	0.03	0.12	(excluded from regression equations)			
5. Lead (Pb)					1.00	0.10	0.36	0.48	0.53	0.33	0.46	0.41	0.35	0.24	-0.08	0.14	0.10	-0.09	-0.11	-0.13
6. Manganese (Mn)						1.00	0.07	0.33	0.57	-0.13	0.65	0.06	0.17	0.42	0.15	0.50	-0.20	0.17	0.09	-0.04
7. Nickel (Ni)							1.00	0.44	0.18	0.76	0.34	0.57	0.45	0.63	-0.21	0.28	0.46	0.07	0.23	0.27
8. Tin (Sn)								1.00	0.45	0.35	0.57	0.42	0.18	0.36	-0.05	0.26	0.05	0.01	0.10	-0.07
9. Titanium (Ti)									1.00	0.02	0.51	0.01	0.05	0.19	0.10	0.30	-0.34	-0.05	-0.27	-0.30
10. Vanadium (V)										1.00	0.10	0.43	0.52	0.64	-0.44	0.13	0.54	-0.03	0.38	0.51
11. Zinc (Zn)											1.00	0.34	0.45	0.47	0.03	0.52	0.12	0.20	0.00	-0.08
12. Nitrogen Dioxide (NO_2)												1.00	0.23	0.46	-0.13	0.37	0.53	-0.20	0.35	0.16
13. Sulphur Dioxide (SO_2)													1.00	0.60	-0.15	0.23	0.47	0.39	0.12	0.47
14. Particulate Sulphate ($SO_4^=$)														1.00	-0.15	0.38	0.45	0.24	0.42	0.51
15. Water Hardness (WH)															1.00	-0.19	-0.44	0.30	0.01	-0.26
16. Arsenic (As)																1.00	0.08	-0.15	-0.26	-0.13

Table 2.6

Comparison of Optimal and Stepwise Regressions for Seventeen Dependent Variables

Same Variables Selected by Both Programs: Number	Number of Cases: Number of Variables in Equation				
	1	2	3	4	5
0	0	2	0	0	0
1	17	3	6	1	0
2		12	1	2	1
3			10	1	4
4				13	3
5					9

rules, the algorithm need not examine every possible combination of N independent variables to find the optimal equation. Instead, the program examines a subset of all possible combinations.

If several of the independent variables are quite good descriptors of the dependent variable, the search will end quickly. If many of the independent variables, or combinations of them, have a similar, small explanatory power, the bounding rules can be relatively inefficient. Consequently, sometimes the search will proceed slowly, and the differences in R^2 will be small. This situation becomes acute for large problems with many independent variables; as the number of variables desired in the equation increases, computation can be excessively time consuming and the results hardly worth the computer time. Four procedures implemented to alleviate this situation are now considered.

First, the user can specify each value of N for which a regression is desired. In most large problems, the level of confidence associated to the F-ratio first increases, and then decreases with N. It is possible, therefore, to set a significance level and do a few, short preliminary searches to find out what values of N should be considered in more depth. For example, suppose there are 25 independent variables. A preliminary job could calculate optimal solutions for N = 5, 10, 15, 20. If the R^2 and F-ratio are sufficiently large at N = 15 but rather small at N = 5, the user may want to examine in more detail values of N between 5 and 15. To use this procedure, see the description of card type 3 in Section 2.7.

Second, the user may limit the number of combinations of variables examined by setting iteration limits (card type 4) on the search for each optimal equation (every N value on card type 3). These iteration limits limit the number of passes through stage 12 of the flowchart; see Figure 2.2.

The rationale for this procedure is that, if the search is long for a given N, the optimal result is probably not much better than numerous other solutions. However, if the iteration limit is reached, the results are no longer guaranteed to be optimal. Another inconvenience of this procedure is that the discrepancies between the actual result and the true optimal result may accumulate over the successive values of N, as the search for the optimal equation involving (N + m) independent variables takes the chosen subset with N variables as a starting point. If the chosen subset of N variables is suboptimal, and if the number of iterations in the search for the optimal subset of (N + m) variables also reaches the iteration limit, the chosen subset of (N + m) variables might be widely suboptimal; therefore, this subset may constitute an inappropriate starting point for the search of the optimal subset for the next equation.

Because suboptimality may become large if several consecutive equations are halted prematurely by the iteration limits, the user can specify that special output be punched by setting IPUNCH = 1 on card type 2. In a later job this punched output can be read into the computer as card types 16 and 17. The search will be continued at the point where it stopped for each equation. The punched output contains the following cards:

1. Unconditional thresholds of variables: punched only once. If the present job uses previous information, these are not punched, since they are the same as before. These form card type 16. The cards described in 2-5 are punched for each regression equation by setting IPUNCH = 1.

2. Values describing problem: the total number of variables, the number corresponding to dependent variable, the value of N and a parameter indicating whether the result was proven optimal or not.

3. MLEVEL values of variables: punched only if not known to be optimal result. These are the integers describing whether the variables are presently in the equation or not and describing in what status they are in the search.

4. The best residual sum of squares so far: punched only if not known to be optimal.

5. MBEST values of variables: always punched. These are integers describing the best subsets of variables so far.

Cards described in 2-5 form card type 17. For more details of this procedure see card types 5, 16 and 17 and the description of NKNØWN on the problem definition card (type 2).

Third, for each regression equation calculated, the maximum likelihood F-ratio is computed. The user can specify confidence limits on the F-ratio for each regression equation. If the F-ratio for a given N falls below its corresponding confidence limit, the program will stop. To use this procedure see the descriptions of card type 6 and ISTØPF on the problem definition card.

Fourth, for each regression coefficient of the N selected variables, the t-test is calculated. The user can specify a confidence limit on t for each regression equation. If the absolute value of t for any coefficient in the regression equation is less than the confidence limit, the coefficient is insignificant. In any one equation, if the number of insignificant coefficients exceeds the maximum set by the user, the program stops. To use this feature, see the descriptions of card type 7 and ISTØPT on the problem definition card. Methods 3 and 4 are optional in any problem since the input required may be time consuming to prepare.

For problems with few independent variables, these stopping methods may not be needed. In large problems of say 25 variables with small values of N(N < 5), the time required is also quite reasonable. Each user's own experience with the program will be the best way to determine the necessity of using these procedures. A preliminary exploration by a stepwise procedure may also be helpful. This can be accomplished with this program by setting each iteration limit equal to one, which simplifies the algorithm to a stepwise regression routine.

2.7 Order and Detailed Description of Input Card Types

The input cards are arranged in the following order:

1.	Title	
2.	Problem definition	
3.	Equations desired	
4.	Iteration limits	
5.	Equations previously processed	(optional)
6.	Confidence limits on F-ratios	(optional)
7.	Confidence limits on t-tests	(optional)
8.	Lower tolerance limits on Durbin-Watson statistic	(optional)
9.	Upper tolerance limits on Durbin-Watson statistic	(optional)
10.	Variables to be forced	(optional)
11.	Format of data	(sometimes can be omitted)

12. Output options
13. Labels for variables (optional)
14. Subset of original variables (optional)
15. Data deck (observations) (sometimes can be omitted)
16. Unconditional thresholds (optional)
17. Information from previous jobs (optional)

Card types 1 through 17 may be repeated in that order any number of times to process several unrelated problems with one execution of the program. This setup is explained with an example in Section 2.8. If several subproblems are to be run from the same data, a format of data card and data deck are needed for the first subproblem only.

2.7.1 Title - FORMAT(20A4)

Any 80 characters may be used to identify the problem.

2.7.2 Problem Definition - FORMAT(13I5,F10.7)

NVAR - total number of variables to be read in, including the dependent variable.

NPØ - number of dependent variable.

NCASES - number of observations.

NMAX - number of equations desired. E.g., if equations with 2 variables, 8 variables, and 15 variables are desired, NMAX = 3.

NFMT - number of cards required to give the format for observation deck. If several different subproblems are desired from the same deck, the data deck need only be read in once. In that case, NFMT = 0 will permit the omission of format cards for subsequent subproblems.

IRES - 0, neither residuals nor the Durbin-Watson d-statistic will be computed; the user does not supply card types 8 and 9 which give tolerance limits on the Durbin-Watson statistic.

Selecting IRES = 0 permits the program to run slightly faster than the four remaining choices for IRES. When IRES > 0, the program must read the observations which it has previously stored on scratch tape (FØRTRAN logical unit 9); the rereading of observations would occur for every equation.

1, residuals will not be printed, but the Durbin-Watson d-statistic is printed; the user does not supply card types 8 and 9, which give tolerance limits for this statistic.

2, residuals will not be printed, but the Durbin-Watson d-statistic is printed; the user must supply tolerance limits for the Durbin-Watson d-statistic by supplying card types 8 and 9.

3, residuals will be printed as well as the Durbin-Watson d-statistic; the user does not supply card types 8 and 9 which give tolerance limits for this statistic.

4, both residuals and the Durbin-Watson d-statistic will be printed; the user must supply card types 8 and 9 giving tolerance limits for the d-statistic.

The remaining options may be unnecessary for many users especially for problems which do not use excessive amounts of time; however, all users should realize that these facilities exist. All of the remaining options can be set equal to zero to turn them off. Consequently, the remainder of the card can be blank for many jobs. The rationale for using options NKNØWN, ISTØPF, ISTØPT, and NSIGN are given in Section 2.6.

NKNOWN - indicates whether parts of the problem have been worked on before, and used chiefly if a previous search was stopped before finding the optimal result. The punched output from this previous job is required as input in this job, if the search is to resume where it stopped.

0, normal, as there is no information from previous runs; therefore card types 5, 16 and 17 will not be included.

i, where i > 0, indicates that information from i equations previously completed or partially completed is to be read in.

A more complete explanation of the purpose is given in Section 2.6, and an example is given in Section 2.7.5.

IPUNCH - 0, normal; no output will be punched.

1, punch card output. This punched output has only one purpose, for use as card types 16 and 17 in subsequent jobs. The output permits a subsequent search in the event the program stops because of the iteration limit, as specified on card type 4, during this job. Also see descriptions of NKNØWN and card types 4, 5, 16, and 17 for the use of punched output.

IFØR - 0, normal.

i, where i > 0, indicates that i variables are to be treated specially; each variable is either to be forced into specified regression equations or to be ignored in determining specified regression equations.

ISTØPF - 0, normal; no confidence limits are supplied by user for F-ratios.

1, confidence limits for each F-ratio are supplied by user; do not stop the program if any equation fails to be significant but print whether equation is significant or not.

2, confidence limits for each F-ratio are supplied by user; should an equation's F-ratio fail to be significant, the program will stop this problem and continue with the next problem (new title card), if possible.

See also card type 6.

ISTØPT - 0, normal; no confidence limits for the t-test are supplied by user.

1, confidence limits for the t-test are supplied by user for each equation; the program prints whether each coefficient is significant or not, but the program does not stop.

2, confidence limits for t are supplied by user for each equation; program prints whether each coefficient is significant or not; program stops this problem if too many coefficients in any one equation are insignificant, and moves to next problem (next title card), if possible.

See also NSIGN and card type 7.

NSIGN - 0, if ISTØPT < 2.

i, where i > 0, if ISTØPT = 2. ISTØPT = 2 is used to stop the program if too many regression coefficients in any one equation are insignificant. When (NSIGN = i) coefficients in one equation are insignificant, the program stops this problem. The program tries to read a new title card to start a new problem, if possible.

See also ISTØPT, card type 7, and Section 2.8. The only case that NSIGN has any effect on the program is when ISTØPT = 2. If ISTØPT = 2, NSIGN must be a positive integer.

II - 0, normal; no technical debugging output will be printed.

1, print detailed intermediate output; this output traces flow through the program and prints the important variable values; intermediate output can become extremely large if there are many variables or if many combinations of variables are tried in a search. THIS IS FOR DEBUGGING PURPOSES AND NOT FOR NORMAL USE.

EPSLØN - blank, normal.

ε, otherwise.

When some independent variables are collinear or nearly collinear, the pivot operation in the correlation matrix to enter or remove a variable from the equation can be undefined because of division by 0. To prevent this, a tolerance level ε is used. If EPSLØN is left blank, the program uses the value 10^{-4}.

2.7.3 Equations Desired - FORMAT(16I5)

Let N be the number of independent variables desired in each optimal regression equation. This card type simply lists the values of N. There must be NMAX of these values, as defined on problem definition card. Specify the values of N in increasing order. For example, suppose for 9 independent variables and the dependent variable, regression equations are desired with 2, 3, 7, and 8 independent variables. Then NVAR = 10, NMAX = 4; the values for the equations desired card are 2,

3, 7, and 8. Values are read into the array, LIST. If NMAX > 16, continue the values of N on a second card.

2.7.4 Iteration Limits - FØRMAT(16I5)

For any regression equation requested, the program requires a limit to the number of iterations performed as a precaution to prevent using excessive amounts of time. Examination of a sequence of combinations of variables for the regression equation is called an iteration; see Figure 2.2. NMAX of these iteration limits are necessary. The first value on the first card corresponds to the first N value on card type 3 (equations desired card). The sixteenth limit on first card corresponds to the sixteenth value of N. If NMAX > 16, continue the iteration limits on a second card. All iteration limits should be greater than zero. Values are read into array NITL.

2.7.5 Equations Previously Processed - FØRMAT(16I5)

For the rationale in using this card, see Section 2.6. This card type is read only when NKNØWN ≠ 0 as defined on the problem definition card. Do not use if NKNØWN = 0.

This card type is used only when partially complete or complete optimal solutions are available from previous jobs. Unless a search had been stopped before the optimal regression equation was found, it is unlikely that this capability would be used. However, it can be used to read in a result that is known to be optimal so that it can be printed with other results, but without recalculating the equation.

The values are integers identifying which equations (which values of N on card type 3) were previously computed so that the program can read in the punched output from the previous run for the appropriate equations. There must be NKNØWN of these values, where NKNØWN was defined on the problem definition card. If NKNØWN > 16, simply continue the values on a second card. The values should be increasing. Values are read into array IKNØWN.

For example, consider a problem with 15 independent variables. On a preliminary, exploratory run, the program found the optimal regression equation with 5 independent variables, but failed to find the optimal solution for 10 independent variables because of the iteration limit. By IPUNCH, the user had requested punched output. Using this punched output, he wants to find optimal solutions with 1, 2, 3, 4, 7, 8, and 10 independent variables. Also, to have all the results together, he would like to read in the answer for 5 independent variables. Therefore, NVAR = 16, NMAX = 8, NKNØWN = 2 (since N = 10 was partially done before and N - 5 was completed before). The N values are 1, 2, 3, 4, 5, 7, 8, 10. The values for this card type are 5 and 10. The program will expect card types 16 and 17 which were punched in an earlier job for N = 5 and N = 10.

2.7.6 Confidence Limits on F-ratios - FØRMAT(8F10.7)

Do not use if ISTØPF = 0. These values provide the confidence limits for the F-ratio of each regression equation. If an equation's F-ratio is less than the confidence limit, the program will print that the result is not significant. If ISTØPF = 2, the program will try to go on to the next problem (next title card). If ISTØPF = 1, it will resume the same problem with the next equation. If the F-ratio is greater than the confidence limit, the program prints that the equation is significant. ISTØPF is defined by user on problem definition card. A confidence limit must be specified for each equation, i.e. for each N specified on the equations desired card. Therefore, NMAX of these confidence limits must be specified. If NMAX > 8, continue the values on a second card. The confidence limits are read into array TF.

2.7.7 Confidence Limits on t-tests - FØRMAT(8F10.7)

Do not use if ISTØPT = 0. These are confidence limits for the t-tests on regression coefficients of independent variables:

$$t = \frac{\text{coefficient of variable}}{\text{std. error of coefficient}} \cdot \left(\frac{1}{1-2/(n-N-1)}\right)^{\frac{1}{2}}$$

where n is the number of observations and N is the number of variables. Confidence limits on t must be specified for each N value on equations desired card if ISTØPT = 1 or 2.

For each equation, the confidence limit specified is compared to the absolute value of t for each coefficient. If the confidence limit is greater than $|t|$, the coefficient is not significant; otherwise, the coefficient is significant. If ISTØPT = 1, the computer prints whether the coefficient was significant or not. If ISTØPT = 2, as well as the printing, the program will stop this problem and move on to the next problem (next title card) if the number of insignificant coefficients for one equation reaches NSIGN, as specified by user on the problem definition card. In some cases the program may not be able to move to the next problem; the program will then stop.

Since a t confidence limit must be specified for each equation, there must be NMAX of these values. These values are read into array STT. Should NMAX be greater than 8, continue the values on the next card.

2.7.8 Lower Tolerance Limits on Durbin-Watson d-statistic - FØRMAT(8F10.7)

Not required when IRES = 0, 1, or 3. Use only for IRES = 2 or 4, where IRES is given by user on problem definition card. For explanation, see next section.

2.7.9 Upper Tolerance Limits on Durbin-Watson d-statistic - FORMAT(8F10.7)

Not required for IRES = 0, 1, or 3. Use only when IRES = 2 or 4. When upper or lower tolerance limits are given by user, the computer calculates the Durbin-Watson d-statistic and compares it to the tolerance limits specified to perform a statistical test. In the following, let DL represent the lower tolerance limit, DU the upper tolerance limit, d the Durbin-Watson statistic:

If $d \leq DL$, the program prints "positive autocorrelation."

If $DL < d < DU$, the program writes "test inconclusive."

If $DU \leq d$, the program prints "random disturbances."

If this capability is used, upper and lower tolerance limits are supplied for each N on equations desired card. Therefore, NMAX of the lower and NMAX upper tolerance limits must be specified. All lower tolerance limits must be specified first; then the upper tolerance limits should be specified on a separate card or cards.

Card types 8 and 9 must be separate. If NMAX = 9, 8 lower tolerance limits should be given on one card and the ninth on the next. Another card must be used to specify the first eight upper tolerance limits, followed by the ninth on a separate card. The lower tolerance limits are read into array DL. Upper tolerance limits are read into array DU.

2.7.10 Variables to Be Forced - FØRMAT(16I5)

Do not use if IFØR = 0, as defined on the problem definition card. Variables may be forced into a particular regression equation by the user, or variables may be specified not to be considered for a particular regression equation. The user must specify which variables are to be treated in this special manner and for which equations these instructions apply.

Each variable which is to be forced into or excluded from an equation has one and only one card of this type. Therefore, there must be IFØR of these cards. The first field of five columns gives the number of the specified variable in columns four and five. The next fifteen fields of five give the information about forcing. To force a variable into an N variable equation, simply place the integer N in the next available I5 field. To exclude a variable from an equation place the integer -N in the next available field. Continue in order of increasing N. If a variable is to be forced into each equation, punch 100 in columns 8-10 of the card.

For example, consider the following three cards and their interpretations:

column:	5	10	15	20
card 1:	3	2	4	-6
card 2:	8	-1	5	
card 3:	10	100		

1. Variable 3 will be forced in the equations of 2 and 4 independent variables, but it will be excluded from the equation of 6 independent variables.
2. Variable 8 will be excluded when choosing one independent variable, but it will be included in the equation for N = 5.
3. Variable 10 will be forced into every regression equation.

The optimal regression program always works with a set of N - 1 independent variables and then chooses variable N by trying all possible solutions. Therefore, the maximum number of variables that can be forced in is N - 1. If NØIN, the number of variables actually in the regression equation, is greater than or equal to N, and if no variables can be pivoted out, the program will print the values of N and NØIN and then stop. This means that N variables cannot be forced into an equation.

Values are read into array KLEVEL. Because of limitations of the size of KLEVEL, the maximum number of variables that can be considered in this option is 15. The maximum number of specific equations that can be given for a single variable is also 15.

Suppose the user has IFØR > 0 and requests punched output. Then, for any equation where variables are forced or excluded by user specification, those variables must be included on card type 10 of any run where that card output is to be read as input.

2.7.11 Format of Data - FØRMAT(20A4)

Do not use if NFMT = 0. A format for the data deck is required whenever observations are read from cards. The data for the regressions can have almost any format, because the program requires the user to specify the format. The preparation of this card follows the rules for FORTRAN IV variable format cards. Consequently, the first character on first card should be a left parenthesis. The total length of the specification cannot exceed 400 characters (5 cards). If the expression is greater than 80 characters, continue on a second card. However, the total number of cards to express the format must be given on the problem definition card under NFMT. The final character of the expression must be a right parenthesis. The expression is stored in array FMT1. As the data itself is read into a double precision array, only F and D type format specifications are allowed. If the data are punched as integer, say as I5, they must be read as real, say as F5.0.

2.7.12 Output Options - FØRMAT(16I5)

This card is required for every problem.

NAPRNT - 0, do not print A matrix.

1, print the A matrix.

The A matrix differs from the covariance matrix by a constant factor. It is defined as follows for the i,jth element:

$$A_{ij} = \sum_{k=1}^{NCASES} (x_{ki} - \bar{x}_i)(x_{kj} - \bar{x}_j)$$

where $\bar{x}_i$ is the mean of the ith variable,

x_{ki} is the kth observation on the ith variable

and NCASES is the number of observations.

NRPRNT - 0, do not print correlation matrix.

1, print correlation matrix.

NMEANP - 0, do not print means of variables.

1, print the mean of each variable.

NØP - 0, do not print observations for this analysis; do not store the observations.

1, print the observations for this analysis; store the observations if NAGAIN = 0.

2, do not print the observations for this analysis; store the observations if NAGAIN = 0.

If IRES > 0 on the problem definition card, the observations will be stored regardless of the value of NØP. Observations need be stored only when (a) residuals or Durbin-Watson statistic are required, (b) output of observation matrix is required, or (c) another problem setup using the same data is required during the same job without repeating the observation matrix.

NAGAIN - 0, the data (observations) will be read from cards; card type 15 is therefore required.

1, instead of reading observations from cards, the program reuses the same observations as the last card deck of observations read. Do not include card type 15.

The option NAGAIN is advantageous if regression equations are to be computed for several subsets of variables selected from a large variable set. Proper use of NAGAIN will permit all the problems to be executed in one use of the program by using card types 1-17 for each problem, but the observation deck (card type 15) need only be included once. See Section 2.8 for a full explanation of the capability and an example; also see card type 14.

If NAGAIN = 1, card type 14 must be included in this problem specification. If NAGAIN = 1, the observations used will be the last set of observations read during this job. When that deck of observations was read, the user must have set NØP = 1 or 2 or IRES = 1, 2, 3, or 4, storing the observations on scratch tape, FØRTRAN logical unit 9. When NAGAIN = 1, the original observations previously read from cards and stored are saved no matter what the value of NØP or IRES. Note that the only way this program is presently coded to read data is from punched cards.

NAMES - 0, no labels for variables.

1, read labels for the variables.

2, use the same labels as the set of labels last read in.

2.7.13 Labels for Variables - FØRMAT(20A4)

Use only when NAMES = 1 on previous card. Labels for every variable are read. The program accepts labels of 8 characters. Spaces are permitted anywhere within the 8-character label. In the output (except for debugging output when II = 1), variables are referred to by their label and relative variable number. Labels are read into array LABELS. A record is kept in array SLABEL.

If there are more than 10 variables, continue the labels on a second card and so on. The program reads labels for each variable; therefore, there must be NVAR of the labels. The first field of eight columns on the first card is the label for variable one; the last field of eight columns on the first card is the label for variable 10.

2.7.14 Subset of Original Variables - FØRMAT(16I5)

Use only if NAGAIN ≠ 0, as defined on output options card. Also, see description of NAGAIN and Section 2.8. This card is used when a set of observations has been read in an earlier problem during the same job. The purpose is to use a subset, or the same set, of the variables that were used in the earlier problem. On this card, the original variable numbers are specified for those variables which are to be used in the current problem. There must be NVAR of these. Values are read into array KVARD. If NVAR > 16, continue on next card.

The computer program must renumber the variables so that the variable numbers of the subset of variables form consecutive integers, the smallest being one. The computer renumbers the variables in the following way. The first variable specified on the subset of original variables card becomes variable one. The sixteenth becomes variable 16. The first variable specified on the second of these cards becomes variable 17. To use this option, the user must have stored the variables on scratch tape in executing the first problem using these data, by setting NØP = 1 or 2, or setting IRES = 1, 2, 3, or 4 during that earlier problem which had NAGAIN = 0.

Renumbering of variables is straightforward as explained above. Note that the renumbering only pertains to the current problem. Suppose observations were read with 10 variables. Let variable 10 be dependent. Suppose all variables are to be candidates for one set of results, and only variables 3, 5, 6, and 9 are to be candidates for a second set of results. Suppose variables 1, 2, 3, 4, 7, and 8 are to be candidates for a third set of results. For the first set of results, NAGAIN = 0 to read the cards (card type 14 is then not included), and set NØP = 2 to store observations. For second set of results, NAGAIN = 1 to use the previous observations without reading in a duplicate card deck. Card type 14 would have:

3 5 6 9 10

For this second set of results. The new variable numbers are:

1 2 3 4 5

To get the third set of results, set NAGAIN = 1. For card type 14, use the original variable numbers, those assigned when the observations were read from cards. So, card type 14 would have:

1 2 3 4 7 8 10

and they would be renumbered:

1 2 3 4 5 6 7

Note when NAGAIN = 1, the previous observations which have already been stored by user are saved no matter what the value of NØP or IRES.

The facility of giving the variables labels was added to the program specifically because the program's renumbering of variables makes output confusing. When using the subset of original variables card, it is wise to use labels. If the labels were assigned by NAMES = 1 when the data were read from cards, then setting NAMES = 2 when a subset of original variables card is read will retain each variable's original name even though its number has been reassigned. This should eliminate confusion in interpreting output. The variable labels are stored as originally read. Hence, one could do a whole regression analysis, followed by an analysis on a subset of the original variables followed by a subset disjoint from the previous one. The variable labels will still be correct as long as NAMES = 2 was specified for each subset.

For example, suppose the data consist of 16 variables with variable 10 as the dependent variable. In the first problem, all variables are to be considered as possibilities for the regression equation. Therefore, NVAR = 16, NPØ = 10, NAGAIN = 0, NØP = 2. However, regressions which exclude certain of the original variables are also to be computed, specifically, a regression with variables 3, 7, 16 as independent variables and variable 10 as the dependent. Then, NØP = 0, NVAR = 4, NPØ = 3, NAGAIN = 1, and the values for the subset of original variables card are 3, 7, 10, 16.

The program was told to use the same data deck as in the first problem; in that problem, NØP = 2 to store the original observations. The variable numbers will be reassigned as follows:

Old	New
3	1
7	2
10	3
16	4

(Example of correct procedure.)

The original observation matrix is retained but only the proper subset of it is used for all computations. At no time should the new variable number be greater than the old variable number; otherwise, the original observation matrix will not be reduced correctly. Suppose:

Old	New
10	1
1	2
4	3
16	4

(Example of incorrect procedure.)

In this case the subset will set the new variables 1 and 2 both to the observed values of original variable 10. Consequently, it is safest to list the old variable numbers on the subset of original variable cards in ascending order.

The number of the dependent variable NPØ has to be specified as the appropriate variable in the new numbering. So, originally NPØ had been 10; but in the example where 3, 7, 10, and 16 were specified in that order, NPØ is 3 because variable 10 is now third. A third, fourth, and fifth problem could be requested, each requiring title, problem definition card and so on; the values of the subset of original variables card should always correspond to the original situation when the data deck was read. The unconditional thresholds are computed after the reduced observation matrix is prepared, as thresholds for a subset will not be the same as for the original set.

2.7.15 Data Deck - User Specified Format

Data are necessary whenever NAGAIN = 0; if NAGAIN = 1, exclude these cards. The program reads the NVAR variables one observation at a time. The data must be organized so that the values of the variables for one observation are read in by one traversal of the variable format card. Hence, the order would be:

$$
\begin{matrix}
x_{1,1}\cdots & x_{1,NVAR} \\
\vdots & \vdots \\
x_{NCASES,1}\cdots & x_{NCASES,NVAR}
\end{matrix}
$$

where NVAR is the total number of variables to be read, NCASES is the number of observations, and x_{ij} is the value of variable j for observation i. The program reads one row at a time.

Values are initially stored in array X, one observation at a time. If the whole matrix needs to be stored, it is written on scratch tape, FORTRAN logical unit 9. Array X is double precision. The observation matrix can be stored by the program on scratch tape only when NAGAIN = 0 by setting NØP = 1 or 2, or IRES = 1, 2, 3, or 4. If a subset of original variables card is used in a problem later in the same execution of the program, only the values for the variables specified on that subset of original variables card are used. However, the whole observation matrix remains stored, so that a second, third or more subset could be specified after the first subset is analyzed by regression analysis. This is explained and an example given in Section 2.8.

2.7.16 Unconditional Thresholds

Also see card types 5 and 17, the description of NKNØWN, and Section 2.6. The unconditional threshold cards were punched by a previous job with IPUNCH = 1. They are necessary only when NKNØWN $\neq$ 0. If NKNØWN = 0, do not include.

1. The first card is a label for the convenience of the user. "THRESHOLDS - O. R. VERSION 6" will be punched in columns 4 - 32.

2. The second card has the unconditional thresholds for variables 1 through 6. Third card has unconditional thresholds for variables 7 through 12, etc. The format is (2X,6D13.5).

The unconditional thresholds are important values calculated for the variables. By their use, the program is able to reduce the total number of combinations of variables examined to a subset of the total. The value of the thresholds is unique to the set of variables and their correlations. If unconditional thresholds were read in by the program, they will not be punched again, even if IPUNCH = 1. In general, if using NAGAIN = 1, the thresholds for the subset of original variables will not be the same as thresholds for original set. See Section 2.7.18 in this regard.

2.7.17 Information from Previous Jobs

Also see card types 5 and 16, the description of NKNØWN, and Section 2.6. These cards were punched by a previous job with IPUNCH = 1. They are necessary only when NKNØWN ≠ 0. Do not include if NKNØWN = 0.

This series of cards must be repeated NKNØWN times, resulting in NKNØWN sets of cards. Each set was punched in the correct order by the computer. As input, each set corresponds to one of the values specified on card type 5, the equations previously processed cards. The first value on equations previously processed card must be the value of N for the first set of card type 17. Similarly, there is a one to one correspondence between the N for each equation on card type 5 and the sets of card type 17. A set consists of the following cards:

1. Identifier - FØRMAT(4I5)

   The first integer is NVAR, the total number of variables including dependent, just as defined on the problem definition card. Second integer is NPØ, the number of the dependent variable, as defined on problem definition card. Third integer is N, the number of independent variables in the regression equation. If any of these three integers is not the same as the present values, the program prints their values, writes that the deck is mixed up, and stops.

   The fourth integer IØPT indicates the situation:

   1, indicates that the result to be read is optimal.

   0, indicates that the result may be suboptimal because the search was stopped when the maximum number of iterations, as specified by the user, was reached.

2. MLEVEL - FØRMAT(16I5)

   These are punched only if IØPT = 0. The array MLEVEL contains integers which indicate whether a variable is presently in the regression equation and at what point the search was stopped. There will be NVAR of these integers.

3. BRSSQ - FØRMAT(D12.6)

   Punched only if IØPT = 0. This is the best residual sum of squares found so far.

4. MBEST - FØRMAT(16I5)

   Always punched; the array MBEST contains NVAR integers indicating which variables gave the best residual sum of squares found yet.

The purpose of these cards is to permit a search for the optimal result to be continued at the very point where it was stopped because of an excessive number of iterations. Also, it permits previous optimal results to be printed in the same output with new results without the program searching again for the optimal solution.

Since the computer punches a set of cards in the correct order, the user need only place the sets in the same order as specified on equations previously processed card. A set was punched for each N in the previous runs. The user must select the N's and corresponding card sets for input.

2.7.18 WARNING!

Combining several of these options can cause difficulties if NKNOWN > 0 and NAGAIN = 1 for the following reason. NAGAIN = 1 means that user wants a subset of the variables from a data matrix which was read earlier. NKNOWN > 0 means that solutions from an earlier use of the program are to be read in, including unconditional thresholds. For a subset of the original variables, the unconditional thresholds that are read must have been computed using precisely the same subset of original variables. Otherwise results will be invalid.

2.8 Several Analyses on Different Sets of Data or the Same Data

Card types 1 through 17 as listed in Section 2.7 form the input to define totally a regression problem. Any number of problems may be done in one execution of the program by setting up another unrelated problem with a second set of the card types. Also, by the appropriate input, several problems may be run on the same set of observations, but considering different variables as dependent.

These capabilities are possible because the program reads the card types as it needs them and because it starts to read a new set of cards only when finished with the previous problem. That is, the program first reads a title card followed by the card types 2-17 as they are needed; some may not be in the card deck since they are optional. When all computations the user requested for this problem are completed, the program attempts to start the next problem by reading a new title card and the other card types. For this new problem, a new set of data may be read. Or, the old set of data can be reused by setting NAGAIN = 1. The program, upon finishing any problem, will attempt to read a new title card, and so on. The process continues until there is no more input to be read; this occurs when a new title card is to be read, if the input is correct. Since the program reaches an end-of-file in input at this time, the computer system will print a warning that this has happened, and the regression program will return control to the computer system. The program may stop before all of the input has been read if:

1. There are too many variables in the regression equation because of user forcing variables.

2. The program is not able to read a title card next. This would occur when equations must still be read in, although the program has stopped the problem because of insignificant results.

3. Card type 17 seems to be incorrect or mixed up.

For every regression problem with a given set of variables, the following card types will always be necessary: title; problem definition; equations desired; iteration limits; output options. If a set of data has been read in earlier in this job, and is to be used again during the same job, a format of data card and data deck will not be needed if the user sets NAGAIN = 1 and NFMT = 0. However, for all other regression problems, a data deck and card(s) to specify its format are required.

For any regression problem, the user specifies NVAR, the total number of variables to be considered in this set of regression equations. Only one of these can be the dependent variable, as designated by NPØ. In some instances, the user may have data on a large set of variables on which he wishes to compute several regression using different variables as dependent. The program can do these analyses in one job if desired, although the procedure is somewhat awkward.

The procedure is to set up a dummy problem to read all of the data as one large observation matrix which will be available for each individual problem. Results of this dummy problem will probably be meaningless. To accomplish this, simply specify any variable as dependent and ask the program to find the optimal solution with N = 1. The program only has to consider one possibility, since the stepwise result for N = 1 is optimal. Consequently, little time is used, but the observation deck is stored in a way that any subset of variables can be considered for the next problem.

To store observation matrix properly, set NØP = 1 or 2, or set IRES = 1, 2, 3, or 4, to read the observations from data cards. Setting NØP = 2 requires the least time; however, the full observation matrix is not printed. Setting NØP = 1 and IRES = 0 is next quickest and prints the observation matrix.

Once the dummy problem has been set up, any regressions may be specified as desired; the following cards are required for each problem: title; problem definition; equations desired; iteration limits; output options; subset of original variables. When subsets of the total set of variables from the dummy problem are specified, the program will internally renumber the variables. Therefore, the user should label the original variables during the dummy problem by setting NAMES = 1 on the output options card and providing labels for variables cards, card type 13, when the observations are read. On subsequent problems, set NAMES = 2 to have the proper labels assigned to the newly renumbered variables. Use of labels should prevent confusion in interpreting output.

Each subsequent problem is defined independently of the dummy problem with the following exceptions. NVAR need not be the same for each problem, though it must be less than or equal to NVAR in the dummy problem. NPØ for all problems after the dummy is the position of the dependent variable on the subset of original variables card. For instance, the dependent variable may be fifteenth in the original data deck but could be fifth on the subset of original variables card for one problem. NPØ = 5 in that problem. More generally, on the subset of original variables card, the variable numbers refer to the original variable numbers when the observations were read. However, NPØ, variable numbers for forced variables, and cards punched by the program in previous uses must correspond to the new numbers of the renumbering system. That is, they must correspond to the position or order as specified on subset of original variables card.

If only one dependent variable is to be used, and if subsets of the original independent variables are to be used, the first problem need not be a dummy problem. It could be meaningful if all the independent variables should be considered for inclusion in the regression equations. In this case, simply specify the original problem as any normal problem.

In summary, in doing several problems, keep these things in mind:

1. If this is the first problem, or if the problem uses a different set of data than the last problem, set NAGAIN = 0, for the data must be read from cards. The data deck and its format must be included. Do not include card type 14, the subset of original variables cards.

2. If reading data from cards by NAGAIN = 0, and if these observations will be used in the next problem or the next i problems, store the observations by setting NØP = 1 or 2 or IRES = 1, 2, 3, or 4. Recall that NØP = 2 will cause the program to run fastest. NØP = 1 simultaneously with IRES = 0 will cause it to run second fastest. Also, give the variables labels. Set NAMES = 1. Do not include card type 14.

3. If the data to be used were the last set read from cards, set NAGAIN = 1 to indicate this. Card type 14, subset of original variables cards, must be included. If labels were given at the time the observations were read from cards, set NAMES = 2 so that the proper variable labels are transferred to the subset of the original variables. Do not include a duplicate card deck of the observations already read. Set NFMT = 0, and do not include a variable format card. The values of NØP and IRES do not affect the observations as stored before, because NAGAIN = 1.

4. Note that only one set of observations can be stored at a time and then only for the duration of that job.

Now consider the following example:

1. Problem: For 4 independent (1, 2, 3, 4) and 2 dependent (5, 6) variables with 50 observations; variables 1, 2 and 5 constitute the first regression problem, and variables 1, 3, 4, and 6 are the second.

2. Dummy problem: NVAR = 6, NPØ = 6, NCASES = 50, NØP = 2, NAGAIN = 0, NAMES = 1. Labels given by user. On equations desired card, N = 1.

3. First problem: To regress variable 5 on variables 1 and 2: NVAR = 3, NPØ = 3, NCASES = 50, NØP = 0 or 1, or 2, NAGAIN = 1, NAMES = 2. Labels will be assigned by program, because NAMES = 2.

   Subset of original variables card: 1 2 5

4. Second problem: To regress variable 6 on variables, 1, 3 and 4: NVAR = 4, NPØ = 4, NCASES = 50, NØP = 0 or 1, or 2, NAGAIN = 1, NAMES = 2. Labels assigned by program, because NAMES = 2.

   Subset of original variables card: 1 3 4 6

Caution. At the present time there are no checks on the range of values specified for the input. For instance, the program would not balk if NVAR $\leq$ 0 or NCASES $\leq$ 0. Of course, this would have no meaning, and could send the program into an infinite loop. Also, one could accidentally specify NPØ > NVAR (which would have no meaning) or NAGAIN = 4 (which has not been defined as a possibility for NAGAIN). Since there are no internal checks that parameters have meaningful values, the user should prepare the input carefully.

The following limitations on problem size should also be observed:

1. 0 < NVAR $\leq$ 40
2. 0 < NPØ $\leq$ NVAR
3. 0 < NCASES $\leq$ 1269
4. 0 < NMAX $\leq$ NVAR - 1
5. 0 $\leq$ NFMT $\leq$ 5
6. 0 $\leq$ NKNØWN $\leq$ NMAX
7. 0 $\leq$ IFØR $\leq$ 15
8. Variable format $\leq$ 400 characters

2.9 Output of the Program

Initially, the program prints the values of the input parameters as specified by user. A few of the output labels need to be explained. "NUMBER OF CASES" is the label for the value of NCASES. The label for the values of N, the input from equations desired card, is labeled "CASES." The confidence limits on t-tests are labeled "STUDENT T TOLERANCES." The values of the card type, variables to be forced, are printed under the headings "VARIABLES TO BE FORCED IN OR OUT" followed by "VARIABLE" and STAGES". The first field of each of those cards is listed under "VARIABLE" while the remaining 15 fields for each card are printed across starting at "STAGES." Fields left blank on the variables to be forced cards will have the value zero printed in the output.

The observation deck, correlation matrix, means of variables, and A matrix will be printed according to the specifications on output options card. During the search for the optimal solution, the following output will be printed for each combination of variables considered:

1. The N independent variables under consideration.
2. If the set gives best results thus far, the program prints the labels and numbers of the N variables, and "WILL BE INTRODUCED." The whole set of N variables will be introduced as the best solution thus far. The old BRSSQ with its R^2 are printed. The new BRSSQ and its R^2 are printed.
3. If the set is not the best so far, the program prints the labels and numbers of the N variables and "WILL NOT BE INTRODUCED." This combination of N variables has been rejected. The residual sum of squares and R^2 of this set are printed. The BRSSQ and its R^2 are printed.

When the optimal solution is found, or when the maximum number of iterations is reached, the program prints the title and value of N. If it is an optimal solution, "OPTIMAL RESULT" is printed. If the search was stopped by the iteration limit, "NONOPTIMAL RESULT" is printed to indicate that the results may not be optimal. Next, the variable names (if specified), numbers identifying variables, and coefficients with their standard errors and t-tests are printed. Also, both the maximum likelihood and the corrected estimate of the following are printed: square of multiple correlation coefficient (R^2), its standard error, multiple R, and F-ratio. (See Section 2.12 for a definition of R^2 corrected for number of observations.) Residuals and Durbin-Watson statistic are printed if desired.

If ISTØPT > 0, the program will print whether or not the coefficients are significant. If ISTØPF > 0, the program prints whether the equation is significant by F-ratio test. Messages will also be printed about Durbin-Watson statistic, if upper and lower limits are supplied by the user.

In the output for a regression equation, it is possible for the program to print for the same result both "OPTIMAL RESULT" and "RESULTS MAY BE NONOPTIMAL." This would mean that results may be nonoptimal because the user specified some variables to be forced, but that the results are optimal within the limitation of having to force variables. "NONOPTIMAL RESULT" and "RESULTS MAY BE NONOPTIMAL", printed simultaneously for an equation, would mean that results may be nonoptimal due to forcing variables, and at the same time that the search was halted by the user-specified iteration limit.

Regarding execution time, the program can be quite slow when compared to stepwise regression programs. Its execution time increases roughly as the square of the number of variables (NVAR). On the IBM 360/65, a FORTRAN H object module executes much faster than a FORTRAN G module; however, a FORTRAN H compilation takes almost twice as long as a FORTRAN G compilation. Values of N and iteration limits should be chosen carefully. In tests conducted using the FORTRAN G compilation, the program uses between 0.1 and 0.2 seconds per iteration to choose, examine, and accept or reject a combination of variables. The program considers many combinations to guarantee optimal results; for this reason the program is slow compared to stepwise procedures. The program can easily be forced to give the same results as stepwise procedures simply by setting the iteration limits at one for each equation (N value) requested. This allows only one combination of variables to be examined.

2.10 Machine Dependent Program Features and Suggestions for Modification

Sections 2.10 and 2.11 are concerned with programming technicalities, and will not usually need to be considered by program users. Section 2.12 concerns a definitional problem in statistics.

The program was written in FORTRAN IV on an IBM 360/65. Unfortunately, the program does use some features of FORTRAN which are unique to the 360. If another computer is used, minor incompatibilities could arise. For instance, all variables declared as REAL*8 would have to be declared as DOUBLE PRECISION. Any variables declared as REAL*4 should be declared as single precision REAL. Variables declared INTEGER*4 should be declared INTEGER.

The READ statement at the beginning of the program has an END parameter. This READ inputs the title card. The END = n means that if an end-of-file is encountered, go to statement number n. For computers other than the IBM 360, appropriate end-of-file procedures need to be substituted. Other incompatibilities may exist; however, they would probably be minor also.

The program assumes a scratch tape is mounted on FORTRAN logical unit nine for storing the data matrix. The user should set up a scratch tape or disk with cards. The program writes in binary on scratch tape for temporary storage. It writes one observation per logical record. Consequently, the logical record length must be at least long enough to store 40 binary double precision variables. If the program is expanded to permit NVAR > 40, the logical record should also be expanded.

The maximum number of observations the program can handle is limited only by the amount of scratch tape storage available. There are no internal limitations on the number of observations. Consequently, the size of the scratch tape and the blocking set up for it are the only limitations on the number of observations, NCASES.

If more than five cards are needed to express the format of the data deck, increase the dimension of the array FMT1 by 20 for each additional format card required. FMT1 is in the main program and subroutine SETUP.

The maximum number of variables can be increased somewhat. To do so, compile the program, and note the amount of core used. Then, from the total amount of core storage available to users, find how much unused core remains. To increase the limit on NVAR to p, eight square matrices with dimension 40 by 40 must be increased to p by p. The arrays are RPRIME, A, R, ØRIG, STORE, ATRIX, B, and C. RPRIME, A, and R are in labeled common and are declared in several subroutines. The other five are local to one subroutine. ØRIG and STØRE are dimensioned in subroutine PIVØTR. ATRIX is dimensioned in RITØUT; and B and C are dimensioned in subroutine PLACE. Remember that each of the arrays is a double precision array, so each element in the array requires two locations. This procedure will permit a new limit p on NVAR $\leq$ p $\leq$ 75.

If there is still unused core and the problems require more than 75 variables, more revisions will be necessary. Suppose the new maximum for NVAR is to be p′, where p′ > 75. The eight square arrays mentioned previously must be increased to p′ by p′. All arrays which have dimension 75 or 80 must be increased to p′. Furthermore, SLABEL, LABELS, and TL must be increased to 2p′. LABELS is a vector in labeled common which is used in several subroutines. SLABEL is a vector local to subroutine SETUP. TL is a vector local to subroutine CHKVAR.

Increasing the limit of the number of variables that can be forced to more than 15 is not recommended. Not only would the array KLEVEL have to be increased to q by 18 for a new limit of q, but many of the DO loop definitions would need a new upper limit of q instead of 15.

2.11 Description of Program by Subroutine

The program consists of a main program and 18 subroutines. The principal arrays and variables not previously defined are documented in this section. Several of the subroutines correspond to stages or boxes in the flowchart, Figure 2.2.

2.11.1 MAIN

MAIN simply controls the flow through the algorithm. Only the simplest of stages are done within this routine; all other steps in the flowchart are separate subroutines. All input card types are read here except for card types 12 through 17. Important arrays include:

- STT - confidence limits for t-test.
- DU - upper tolerance limits on Durbin-Watson statistic.
- DL - lower tolerance limits on Durbin-Watson statistic.
- IKNØWN - values from card type 5; these are the values of N for equations previously processed.
- TF - confidence limits for F-ratio.
- LIST - values of N read from card type 3 (equations desired card).
- NITL - limits on number of iterations in search for optimal solution; read from card type 4.
- FMT1 - format of data; read from card type 11.
- TITLE - title of problem; read from card type 1.

KLEVEL - first column of the 15 by 18 array contains numbers of variables to be forced; columns 2 through 16 contain information as to N values where variables are to be forced; column 17 lists the variables to be kept out of equation at the N value being worked on; column 18 lists variables to be in equation at N value under consideration.

Important variables not defined in input are:

IREAD - binary indicator:

0, no information from earlier jobs for this equation.

1, information from earlier job was read for this equation.

N4 - binary indicator:

0, information read in for this equation may be suboptimal.

1, information from previous job is optimal for this N.

MLIST - pointer to keep track of what elements should be used in arrays LIST, NITL, TF, DL, DU, STT.

NITE - number of iterations for this N.

MØW - limit on number of iterations for this N.

BRSSQ - best residual sum of squares found for combinations of variables tried so far.

IFØUND - variable chosen at stage 4.

MAXØD - maximum odd value of elements in array MLEVEL; if this is negative, MAXØD = 1; when MAXØD = 1, a solution should be printed.

NØIN - number of independent variables presently pivoted in.

N - number of independent variables specified for the regression equation.

IMULRN - binary indicator:

0, normal.

1, one of the tests for insignificance has caused the program to move to a new problem (new title card).

NPNVAR - NVAR when a data deck was last read; if it is first problem, NPNVAR = 1 initially.

MAIN also controls exits from the program; completion of every problem specified occurs here.

2.11.2 SETUP

SETUP computes the covariance matrix, means, and correlation matrix. The observation matrix is read in, but is stored on disk only if requested or if a table of observations, residuals, or Durbin-Watson statistic is requested. The output options card is read followed by card types 13, 14, and 15 if they are to be used in this problem.

Important new arrays used in this subroutine are:

X - temporary storage array for data, as the observations are read one at a time.

A - proportional to the covariance matrix; for a definition, see Section 2.7.12. After the correlation matrix is computed, this matrix is destroyed, as it is used as a work array.

R - correlation matrix.

ADIAG - a one dimensional array for storing the diagonal elements of the A-matrix before it is destroyed; it contains the sums of squared deviations about the variable means.

XSUM - means of the variables.

MLEVEL - designates which variables are presently in the regression equation and the status of the search.

MBEST - indicates which variables are in the best solution thus far.

KVARD - subset of original variables to be considered in this problem; read from card type 14.

LABELS - variable labels in the present problem.

SLABEL - labels of the variables the last time that labels were read in from card type 13; this storage array permits old labels to be used when a subset of original variables is used.

If previous data from the same job are to be used in another problem, card type 14 is read. The data in this case will be read from scratch tape, unit 9. The variables are renumbered, and the observations read from the scratch tape, but only values for the specified variables are used. Then means, A matrix, and R matrix are computed. Labels are read and assigned if used. NPNVAR is updated if NAGAIN = 0.

If some of the independent variables are dummy variables, all column and row elements of correlation matrix are set equal to zero for the dummy variables with the following exceptions: the diagonal element is set equal to one and the correlations between dummy variables are set equal to one. All elements not involving a dummy variable are calculated as usual. The above procedures detect dummy variables by checking the product of diagonal elements of A-matrix for each pair of variables.

If the product is less than 10^{-25}, the corresponding element of correlation matrix is zeroed. MLEVEL is initialized to -1 for all independent variables.

2.11.3 STAGE2

STAGE2 computes the unconditional thresholds of the independent variables. Unconditional thresholds are the residual sum of squares with all linear independent variables in the regression equation except the variable whose threshold is being computed.

If NKNØWN > 0 as specified by user, the program reads card type 16, the unconditional thresholds, and prints them six per line. The subroutine exits immediately; consequently, even if IPUNCH = 1, the unconditional thresholds will not be punched again. If NKNØWN = 0, the subroutine computes unconditional thresholds. If IPUNCH = 1, the thresholds are punched six per card. This punched output is card type 16 to be used for future runs.

STAGE2 calls PRIMR which pivots in linearly independent variables. Important arrays for STAGE2 are:

T - unconditional thresholds.

RPRIME - work array containing the correlation matrix stored in A, but with the variable, whose threshold is being computed, pivoted out. Note that the correlation matrix as stored in A is really the resultant correlation matrix with all linearly independent variables pivoted in.

IMP - same as in PRIMR, which follows.

A - same as in PRIMR.

All other arrays and variables are as defined earlier.

2.11.4 PRIMR

PRIMR is only called from subroutine STAGE2. Its purpose is to compute the correlation matrix with all linearly independent variables pivoted in the equation. It remembers which variables did not pass the epsilon tolerance test to be pivoted in, i.e., those variables which have linear dependencies with variables already pivoted in by PRIMR.

Important arrays are:

IMP - contains the number of each variable which failed the tolerance test. If there were two variables which could not pass the tolerance level, say variables 5 and 7, then IMP(1) = 5, IMP(2) = 7, and for all I $\neq$ 1 or 2, IMP(I) = 0.

A - resultant correlation matrix with all variables pivoted in that passed the epsilon tolerance.

2.11.5 STAGE4

STAGE4 chooses the variable that gives the largest reduction of the residual sum of squares. Therefore, for all variables I with MLEVEL(I) = -1, the subroutine finds the single variable for which $R(I,NP\emptyset)^2/R(I,I)$ is a maximum, given that R(I,I) > EPSLØN. The number of the variable is stored in IFØUND for future use.

2.11.6 CHKVAR

CHKVAR executes stages 7, 9 and 10 of the flowchart (Figure 2.2). The best residual sum of squares to date is located in BRSSQ. The residual sum of squares with the combination of variables under consideration, RSSQ, is computed as

$$\left(R(NP\emptyset,NP\emptyset) - \frac{R(IF\emptyset UND,NP\emptyset)^2}{R(IF\emptyset UND,IF\emptyset UND)} \right) A(NP\emptyset,NP\emptyset)$$

where IFØUND is the same as in stage 4, and A(NPØ,NPØ) is the (NPØ,NPØ) element of A. The value of A(NPØ,NPØ) is located in ADIAG(NPØ).

The program next prints out the variables that are presently being considered for the regression equation. If the variables have labels, the labels are printed, as well as the variable numbers. Otherwise, only the numbers of the variables are printed.

If RSSQ, the residual sum of squares is greater than or equal to BRSSQ, the program prints that the variable will not be introduced. This entire combination of variables has been rejected. The residual sum of squares and R^2 with this combination is printed. The best residual sum of squares is printed and best R^2 to date is printed. The program then exits from CHKVAR.

If the computed residual sum of squares is less than BRSSQ, the program introduces this combination of variables as the best thus far. The new BRSSQ is printed with its R^2; the old BRSSQ and R^2 are printed. The program then executes stages 9 and 10.

The important arrays are:

LVP - numbers of the variables presently in the regression equation.

MBEST - array designating which variables are in the best set of regression variables for this value of N:

MBEST(I) $\geq$ 0 means that variable I is in the best set so far, unless I is a variable forced to be ignored.

MBEST(I) < 0 means that I is not in the best set.

MLEVEL - array designating whether variables have been pivoted into the regression equation:

MLEVEL(I) $\geq$ 0 means variable I has been pivoted in, unless I is a variable forced to be ignored.

MLEVEL(I) < 0 means I has not been pivoted in.

T - unconditional thresholds.

Recall that KLEVEL(I,17) lists the variables which the user specified to be ignored for this N. KLEVEL(I,18) lists variables which the user wants forced in for this N.

For stage 9, the work is straightforward. In stage 10, the search is conducted to find any variables which are good enough to stay in any future best set of regression variables, based on the unconditional threshold. Those that pass the test are given MLEVEL = 0.

2.11.7 STAGE8

STAGE8 simply pivots in variables chosen at stage 4. It is only used when the number of variables already pivoted in is less than N - 1.

The only important new variables are:

NOIN - number of variables that are currently pivoted in.

MAXOD - maximum odd value of MLEVEL.

2.11.8 FMAXØD

Determines the value of MAXØD, the maximum odd MLEVEL. If the maximum is less than zero, MAXØD is set equal to one.

2.11.9 ØUTAT1

ØUTAT1 performs stage 16, which is relatively straightforward. The variables that have MLEVEL $\leq$ -MAXØD are already out of the equation; therefore, only each

MLEVEL has to be revised. Every variable with MLEVEL > MAXØD is pivoted out of the equation, and its MLEVEL is set to -1. NØIN is updated.

2.11.10 ØUTATM

This subroutine performs stage 17. All of the variables are the same as in other subroutines.

Of the variables J, satisfying MLEVEL(J) = MAXØD, the variable is chosen which maximizes $R(J,NPØ)^2/R(J,J)$. This variable is pivoted out and its MLEVEL set to minus MAXØD.

2.11.11 CØNDTH

This is stage 18, which computes the conditional thresholds for variables with MLEVEL = -1. The conditional threshold is the residual sum of squares with all variables pivoted in except:

1. the variable whose threshold is being computed,
2. variables with (- MAXØD -1) < MLEVEL < (-1), and
3. variables that do not pass the tolerance test when the program tries to pivot them in.

As in the procedure used in stage 2, the subroutine pivots all variables desired into the equation and stores the correlation matrix in the array A. Then, as the conditional threshold for each variable is computed, the subroutine pivots that variable out and stores the correlation matrix in RPRIME. An exception occurs if the variable had not been pivoted in originally because it failed the tolerance test. Since the variable was not pivoted in, instead of using the values in matrix RPRIME, matrix A is used. Variable NYES is a binary indicator to indicate whether the variable whose conditional threshold is to be computed is such a variable. NYES = 1 means it is; NYES = 0 means that it is a normal variable and must be pivoted out. If the conditional threshold of the variable is greater than or equal to the best residual sum of squares, the variable is selected. The procedure is continued until either all variables are checked or the number of variables in the regression equation is one less than the number desired. Therefore, it is possible that the subroutine will return before the threshold for each variable with MLEVEL = -1 has been computed.

All thresholds that have not been computed are set equal to zero.

The array CT contains the conditional thresholds. The array IMP is used the same way as in PRIMR and STAGE2. RPRIME is a work array used in computing the conditional thresholds.

2.11.12 PLACE(B,C)

This subroutine transfers elements of array B to array C.

2.11.13 PIVOTR(ØRIG,STØRE,BG,INVRNØ,IDIM,IDØNE)

This subroutine performs all pivot operations for placing a variable in the regression equation or removing it from the equation. The arguments are:

ØRIG - input matrix for the pivot operation.

STØRE - return matrix which will contain the result of the pivot operation. STØRE and ØRIG may be the same array.

NG - variable to be pivoted in or out.

INVRNØ - 0, means pivot variable NG in.

1, means pivot variable NG out.

IDIM - dimension of square matrices ØRIG and STØRE.

IDØNE - return variable;

if 1, pivot operation could not be performed because of the epsilon tolerance test.

if 0, pivot was performed.

A variable that has been pivoted in can always be pivoted out; so, the epsilon test is only used when pivoting a variable in.

2.11.14 RITØUT(ATRIX,INØUT,NEL)

The subroutine prints the results of a pivot operation. If II = 1 (if no intermediate checking output is desired), the subroutine returns without any printing. The contents of ATRIX are printed one complete row at a time. Ten elements are printed per line. Each new row starts on a new line. The contents of ATRIX are printed no matter what the value of INØUT.

ATRIX - matrix to be printed.

INØUT - 0, prints ATRIX and that variable NEL was pivoted in.

1, prints ATRIX and that NEL was pivoted out.

>1, prints contents of ATRIX, but no messages.

NEL - variable that was pivoted.

(Note that the II during the program is actually 1 - II', where II' was given by user on problem definition card.)

2.11.15 RESET

This subroutine performs stage 15. However, the next value for N is assigned in the main program.

2.11.16 ØUTPUT

ØUTPUT performs stage 13, which prints the final results: title, number of variables, variables in the equation, coefficients and their standard errors, constant, maximum likelihood estimate of R^2 and R, standard error on R^2, maximum likelihood F-ratio, corrected estimate of R^2 and R, standard error on corrected R^2, and corrected F-ratio.

The first step is to perform any pivot operations necessary to place the optimal variables in the equation. If the subroutine has been called from stage 4A, the subroutine immediately returns to stage 4A. The signal for this return is N4 = 0. If the subroutine was called from the main program, N4 = 1; and the subroutine prints the regression equation and statistical output. Note several things about these pivot operations. The pivot operations in ØUTPUT assume that forced variables do not have MLEVEL = 0 nor MBEST = 0. Also, MLEVEL is not updated to correspond to these pivot operations. Hence, after these operations have all been completed, MBEST, not MLEVEL, tells what variables are currently selected. If ØUTPUT was called by the MAIN program, subroutine RESET (stage 15) will update MLEVEL to correspond to the variables currently pivoted in. If ØUTPUT was called by subroutine STG4A, then STG4A will do it.

Residuals and/or Durbin-Watson statistic can be computed if the user desires. However, this requires the reading of the observations from scratch tape; therefore, the computation is time consuming. If the stopping procedures indicated by ISTØPF > 0 and/or ISTØPT > 0 are to be used, the checking will be done here. The value of N is the number of independent variables in the regression equation.

New arrays are as follows:

CØEF - stores the coefficients.

Ø - stores the observed value of dependent variable.

P - stores the value of dependent variable as predicted by equation.

RSD - residuals.

IA - number of the observation.

Since the residuals are printed out three per line, Ø, P, RSD, and IA are only three locations each in length.

STT - confidence limits on t-test.

DU - upper tolerance limits on Durbin-Watson statistic.

DL - lower tolerance limits on Durbin-Watson statistic.

X - array into which observations are read one at a time.

TL - confidence limits on F-ratio.

KLEVEL(I,17) - variables to be excluded from this equation; all entries after end of list must be zero.

KLEVEL(I,18) - variables forced into equation; all entries after end of list must be zero.

Important variables are:

DW - Durbin-Watson statistic.

AR - maximum likelihood estimate of R^2.

BETAØ - constant for regression equation.

ARS - maximum likelihood R.

F - F-ratio, used for both maximum likelihood and corrected estimates.

BR - corrected estimate of R^2.

BRS - corrected estimate of R.

NITE - number of iterations tried so far for equation.

MØW -- limit on iterations.

IMULRN - binary indicator:

0, normal.

1, stop this problem; move to next title card, if possible.

The search was halted before completion if NITE equals MØW.

2.11.17 STG13A

This subroutine performs tasks which are really part of stage 13; therefore, this subroutine does not appear as a separate entity on the flowchart. Though it should be considered part of stage 13 on the flowchart, it is separate from ØUTPUT, the subroutine which performs most of stage 13. STG13A is called from the main program immediately before the main program calls ØUTPUT.

The subroutine has two purposes. If variables have been forced (that is, if user specified IFØR > 0), the forced variables are prepared for the output. Those which were forced into the equation have their MBEST and MLEVEL set to three. Those which were specified to be ignored have their MLEVEL and MBEST set to minus one.

If punched output was requested, it is done here. Card type 17 is punched here in the order described in Section 2.7.17.

Note again variables forced into the regression equation are listed in KLEVEL(I,18). All entries after the end of the list must be zero. Variables purposely excluded are listed in KLEVEL(I,17). All entries after the end of this list must also be zero.

2.11.18 STG4A

STG4A has two purposes: to do all initializations if some variables are to be forced, and to read input from previous runs. If NKNØWN > 0, the user has specified equations on card type 17 to be read. The program first checks to see if this value of N is one for which there is previous information on cards. If not, the program goes to the section for forcing variables, having set N4 = 1 and IREAD = 0.

If the N is one for which card information is provided, the program reads card type 17. If the previous result may not be optimal, the program calls ØUTPUT immediately after reading the MLEVEL's into array MBEST; upon returning from ØUTPUT, MLEVEL is set equal to MBEST. If the previous result is optimal, none of this occurs. The true values of MBEST are read into MBEST as the final part of card type 17 whether the previous result is optimal or not. MAXØD is set to one, and IREAD is set to one. Should the cards read not be for this problem or equation, the program prints an error message and stops.

The coding to force variables is next. If IFØR = 0, this coding is skipped. First, the program searches the array KLEVEL to find the variables to be treated specially for this N. As it finds them, it performs several operations:

1. If a variable is to be excluded, the number of the variable is stored in column 17 of matrix KLEVEL. If this variable is in the equation, it is pivoted out and NØIN is updated.
2. If a variable is to be forced into the equation, its number is stored in column 18 of matrix KLEVEL. The variable is pivoted into the equation and NØIN is updated, if the variable is not already in the equation.
3. For all variables to be forced either in or out of equation, MBEST and MLEVEL are set to zero. This has the effect that the variable will not be pivoted any more. STG13A restores the variables to normal when the solution is found.

If card type 17 was read for this N (i.e. if IREAD = 1), subroutine STG4AB is not called. However, if card type 17 was not read, then if NØIN $\geq$ N or variables have been forced either in or out for this particular value of N, subroutine STG4AB is called.

Array KLEVEL contains all information about variables to be treated specially. Array IKNØWN contains the value of N for which information was previously obtained and put on cards. Important variables are:

N4 - binary indicator:

0, equation read may not be optimal.

1, equation read is optimal.

IREAD - binary indicator:

0, no equation read.

1, an equation was read.

All other variables and arrays are defined as in several other subroutines. As before, column 17 of KLEVEL lists variables user has specified to be excluded. All entries after end of list must be zero. Column 18 of KLEVEL lists variables user has forced into the equation. All entries after end of list must be zero.

2.11.19 <u>STG4AB</u>

This subroutine is only called from STG4A in certain instances explained above. When called, variables have been forced for this equation, or NØIN $\geq$ N; therefore

BRSSQ must be updated.

It is possible, though unlikely, that by forcing several variables, NØIN could be greater than N - 1. Since the program always works with a set of N - 1 independent variables in the equation, variables must be pivoted out. If NØIN $\geq$ N, the program pivots out of the equation that variable(but not a forced variable) which would least increase the best residual sum of squares. That is, it maximizes $R(I,NPO)^2/R(I,I)$, where I is the number of an independent variable in the equation which is not forced. Since variable I is in the equation, $R(I,I) < 0$. The MLEVEL and MBEST of the variable which least increases BRSSQ are set to minus one, the variable is pivoted out, and NØIN is updated. If no such variable can be found, that is, if more than N - 1 variables were forced into equation, the program prints that the situation is hopeless; then the program stops. If a variable was found, BRSSQ is updated and the procedure is repeated by checking to see if NØIN $\geq$ N.

Since BRSSQ has been updated (and perhaps increased because of forced variables), all nonforced variables with MLEVEL = 0 must be checked to make sure their unconditional threshold is still greater than or equal to BRSSQ. If a nonforced variable with MLEVEL = 0 has its unconditional threshold less than BRSSQ, the program resets the variable's MLEVEL and MBEST to 3. All of this is done just before returning to STG4A, after BRSSQ has been updated the last time in STG4AB.

2.12 Definitions of the Square of the Multiple Correlation Coefficient

The optimal regression program computes two related estimates of the square of the multiple correlation coefficient, ρ^2. One is based on the usual definition, and the second is adjusted for the number of degrees of freedom. This note sets forth the definitions of each and discusses their properties.

Consider n observations on p+1 random variables, x_{ij}, i = 1...n; j = 1...p+1. Let Σ and R be the covariance and correlation matrices associated with the variables.

$$\Sigma = \begin{pmatrix} \Sigma_{11} & \Sigma_{12} \\ \Sigma_{21} & \Sigma_{22} \end{pmatrix} \begin{matrix} p \\ 1 \end{matrix} \qquad R = \begin{pmatrix} R_{11} & R_{12} \\ R_{21} & R_{22} \end{pmatrix} \begin{matrix} p \\ 1 \end{matrix}$$

$$\begin{matrix} \quad p & \quad 1 \end{matrix} \qquad\qquad \begin{matrix} \quad p & \quad 1 \end{matrix}$$

The usual definition of the square of the multiple correlation coefficient ρ^2 between variable p+1 and the first p variables is:

$$\rho^2 = \frac{\sum_{21}\sum_{11}^{-1}\sum_{12}}{\sum_{22}} = R_{22}\, R_{11}^{-1}\, R_{12}$$

Let $S = (s_{jk})$ be the usual estimate of $\sum$, where

$$s_{jk} = \frac{1}{n-1}\left(\sum_{i=1}^{n}(x_{ij} - \bar{x}_j)(x_{ik} - \bar{x}_k)\right)$$

The usual estimate R^2 of ρ^2 is

$$R^2 = \frac{S_{21}\, S_{11}^{-1}\, S_{12}}{S_{22}}$$

This is the estimate adopted in the optimal regression program.

Now consider the regression of variable p+1 on the first p variables as the regression of a random variable on p nonrandom variables; the F-test of the hypothesis $\rho = 0$ involves the conditional distribution of R^2, given $x_1 \ldots x_p$.

$$F = \frac{R^2}{(1 - R^2)} \cdot \frac{(n-p-1)}{p}$$

has an $F_{p,n-p-1}$ distribution, that is, the variance-ratio distribution with p and n-p-1 degrees of freedom. This distribution is central when ρ^2, the true value of the square of the multiple correlation coefficient, is zero, and noncentral with a noncentrality parameter $\lambda = n\rho^2$ when its true value is ρ^2.

However, to consider the set of independent variables as nonrandom is incompatible with the use of the optimal regression program or a stepwise regression program. Indeed, to consider the set of independent variables as fixed implies the assumption, according to reasons concerning the nature of the problem and variables, that the dependent variable is a linear combination of <u>all</u> the independent variables plus a random error term. This is incompatible with the consideration of subsets of independent variables, as is the case in both the optimal and stepwise regression programs. Furthermore, to consider the independent variables as fixed would prevent any use of the results for prediction. Thus, the results would only be valid for the sample and not for a wider population.

For these reasons, now assume the independent variables are also random, which implies the observations should be drawn at random, which is not generally the case. However, this should be done whenever possible. Now consider the distribution of R^2 under this assumption. If the true value of the square of the multiple correlation coefficient ρ^2 is zero, then

$$F = \frac{R^2}{(1 - R^2)} \cdot \frac{(n-p-1)}{p}$$

has a central $F_{p,n-p-1}$ distribution. Thus, the test of significance of the regression equation is the same as when the independent variables are considered fixed. This is the test used in the optimal regression program.

If the true value of the square of the multiple correlation coefficient, ρ^2, is not 0, the distribution of R^2 is much more complicated as shown by Anderson (1958). However, the true value of ρ^2 cannot be considered equal to zero whenever the regression equation is significant. In this case, the expectation of R^2 is, approximately

$$E(R^2) = \rho^2 + \frac{p}{n-1}(1 - \rho^2) + \frac{2(n-p-1)}{(n^2-1)}(1 - \rho^2)\rho^2$$

and the variance of R^2 is approximately

$$\frac{4\rho^2(1 - \rho^2)(n-p-1)^2}{(n^2-1)(n+3)}$$

Therefore, R^2 as defined above is a heavily biased estimate.

Olkin and Pratt (1958) have derived an unbiased estimate $\hat{R}^2$ of ρ^2.

$$\hat{R}^2 = R^2 - \frac{n-3}{n-p-1}(1 - R^2) - \frac{2(n-3)(1 - R^2)^2}{(n-p-1)(n-p+1)}$$

Other estimates also can be used, reducing the bias without completely eliminating it. The one used in the typical stepwise regression program is the following:

$$\hat{\hat{R}}^2 = \frac{n-1}{n-p-1} R^2 - \frac{p}{n-p-1}$$

It is therefore related to R^2 by a simple formula. By computing $E(\hat{\hat{R}}^2)$ it is possible to see to what extent this estimate reduces the bias. Using the approximation formed by dropping the last term from the above expectation of ρ^2:

$$E(R^2) = \rho^2 + \frac{p}{n-1}(1 - \rho^2) \text{ , then}$$

$$E(\hat{\hat{R}}^2) = \rho^2 \frac{n(n-p-1) - (n-1)}{(n-p-1)^2}$$

In this case, if the true value of the correlation coefficient is zero, so is the approximate expectation of $\hat{\hat{R}}^2$. In this sense, $\hat{\hat{R}}^2$ can be said to reduce the bias; $\hat{\hat{R}}^2$ can be considered to be deduced from R^2 by the following procedure. Recall that the residual variance is

$$\Sigma_{22}(1 - \rho^2) = \Sigma_{22}\left(1 - \frac{\Sigma_{21}\Sigma_{11}^{-1}\Sigma_{12}}{\Sigma_{22}}\right)$$

The estimate of this residual variance in the optimal regression program is

$$S_{22}(1 - R^2) = S_{22}\left(1 - \frac{S_{21}\, S_{11}^{-1}\, S_{12}}{S_{22}}\right)$$

In the typical stepwise program, it is

$$S_{22}(1 - \hat{\hat{R}}^2) = \frac{n-1}{n-p-1} S_{22}\left(1 - \frac{S_{21}\, S_{11}^{-1}\, S_{12}}{S_{22}}\right)$$

Thus, $\hat{\hat{R}}^2$ is based on a correction of the residual variance for the degrees of freedom. This corrected residual variance is used to compute the "corrected" R^2 and F-ratio in the optimal regression program. As noted in Section 2.3.1, use of the uncorrected residual variance in the algorithm does not affect the optimal selection of variables.

3. INTERDEPENDENCE ANALYSIS

3.1 Introduction

In data analysis, many situations arise in which one wishes to select a representative subset of variables from a larger set. By representative, one generally means that the selected subset best describes or explains the entire set in some well-defined sense. Beale, Kendall and Mann (1967) proposed a procedure, which they called interdependence analysis, for accomplishing this objective.

Interdependence analysis belongs to a general set of procedures for replacing a large set of variables with a smaller set. Principal components analysis attacks this objective by specifying a new set of variables (i.e. components) which is a linear combination of the original variables; see Morrison (1967). Alternatively, factor analysis seeks to solve the same problem by specifying a new set of variables, (i.e. factors) of which the observed variables are linear combinations. A somewhat related technique is the automatic interaction detector of Sonquist and Morgan (1970). Their procedure seeks to account for the variance of a given dependent variable by partitioning the sample of observations. Many candidate variables or factors are used to define the levels in a sequence of one-way analyses of variance. Those factors chosen from the entire set as best partitioning the sample may be viewed as a subset selected from a larger set.

The algorithm for interdependence analysis described in this chapter enables N variables to be chosen from a larger set of p variables. By selecting variables, rather than creating new ones, this procedure avoids the perplexing problem of interpreting the principal components or factors. The criterion for choosing the variables is straightforward. Consider each possible subset of N variables as the independent variables in a regression equation with each of the (p - N) "rejected" variables as the dependent variable; this results in (p - N) regression equations, each with its R^2, the estimate of the square of the multiple correlation coefficient. Then, select the subset of N variables that maximizes the minimum of these (p - N) values of R^2.

The algorithm of Beale, Kendall and Mann finds this optimal solution without checking the $\binom{p}{N}$ combinations of N variables. Bounding rules are used to limit the search to a subset of the total number of combinations. Even with the bounding rules, the program checks many combinations; consequently, it can be relatively slow.

The contents of this chapter closely follow the organization of Chapter 2. The algorithm is described in Section 3.2 together with definitions of the pivot operations, which are largely repeated from Chapter 2 for the convenience of the reader. Section 3.3 describes an example of the application of the program to the same set of independent variables used in the example of Chapter 2. Strategies in using the program are outlined in Section 3.4, followed by detailed descriptions of input card types, program output, machine dependent features and subroutines.

3.2 Interdependence Analysis Algorithm

The interdependence analysis algorithm is another application of the general algorithm for selecting subsets of variables (Figure 2.1), and therefore closely resembles the optimal regression algorithm of Figure 2.2. Although the basic computation equations are only slightly different than those described in Section 2.2.2, they are repeated here for clarity and convenience of the reader. As this application of the algorithm is less familiar than the regression case, a more detailed explanation is provided.

3.2.1 Pivot Operations

Let X be a data matrix of n observations on p variables:

$$X = \begin{pmatrix} x_{11} & \cdot & \cdot & x_{1,p-1} & x_{1,p} \\ \cdot & & & & \\ \cdot & & & & \\ \cdot & & & & \\ x_{n1} & \cdot & \cdot & x_{n,p-1} & x_{n,p} \end{pmatrix}$$

Let $A = (a_{jk})$ be the p by p matrix of sums of squares and cross products, and $R = (r_{jk})$ be the p by p sample correlation matrix. The following operations are now defined on the sample correlation matrix:

1. The transformation P_q on the matrix R (q = 1...p) is called "pivoting in the variable q": $R' = P_q(R)$

$$r'_{qq} = -1/r_{qq}$$

$$r'_{qj} = r'_{jq} = -r_{qj}/r_{qq}$$

$$r'_{jk} = r'_{kj} = r_{jk} - \frac{r_{jq}r_{qk}}{r_{qq}}$$

This transformation is not defined when $r_{qq} = 0$.

2. The transformation Q_q on the matrix R (q = 1...p) is called "pivoting out the variable q": $R' = Q_q(R)$

$$r'_{qq} = -1/r_{qq}$$

$$r'_{qj} = r'_{jq} = r_{jq}/r_{qq}$$

$$r'_{jk} = r'_{kj} = r_{jk} - \frac{r_{jq}r_{qk}}{r_{qq}}$$

The successive application of the two transformations P_q and Q_q on R results in the original matrix R: $Q_q(P_q(R)) = R$. To obtain the regression equation of x_k on say $x_{\alpha 1} \dots x_{\alpha N}$, it is necessary to pivot in successively the variables $x_{\alpha 1} \dots x_{\alpha N}$. Let R′ be the resulting matrix, $R' = P_{\alpha N} \dots P_{\alpha 1}(R)$. To avoid the difficulties of near collinearities, if in this pivoting process a pivot element r_{qq} is smaller than a given level of tolerance ε, the corresponding variable will not be pivoted in.

The following standard definitions are relevant to understanding the algorithm:

1. Estimate of square of the multiple correlation coefficient: $R^2 = 1 - r'_{kk}$ for a regression equation for variable k on N variables pivoted in as independent variables. Note that variable k is not pivoted in.

2. Estimate of standard error on R^2:

$$\frac{\left[(4R^2)(1-R^2)(n-N-1)^2\right]^{\frac{1}{2}}}{\left[(n^2-1)(n+3)\right]^{\frac{1}{2}}}$$

3. Estimates of regression coefficients for x_k on $x_{\alpha 1} \dots x_{\alpha N}$ are:

$$\beta_{\alpha 1} = -r'_{\alpha 1,k} \frac{(a_{k,k})^{\frac{1}{2}}}{(a_{\alpha 1,\alpha 1})^{\frac{1}{2}}}$$

$$\vdots$$

$$\beta_{\alpha N} = -r'_{\alpha N,k} \frac{(a_{k,k})^{\frac{1}{2}}}{(a_{\alpha N,\alpha N})^{\frac{1}{2}}}$$

$$\beta_o = \bar{x}_k - \beta_{\alpha_1}\bar{x}_{\alpha_1} \dots - \beta_{\alpha N}\bar{x}_{\alpha N}$$

where $\bar{x}_i$ is the mean of the variable x_i.

4. Estimate of standard error on $\beta_{\alpha q}$ ($q = 1 \dots N$) is given by:

$$\frac{a_{k,k}}{a_{\alpha q,\alpha q}} \cdot \frac{1-R^2}{1-S^2_{\alpha q}} \cdot \frac{1}{(n-N-3)}$$

where $(1-S^2_{\alpha q})$ is the (α_q,α_q) element of the matrix derived from R' by pivoting out the variable $x_{\alpha q}$.

5. F-ratio is equal to:

$$F = \frac{n-N-1}{N} \cdot \frac{R^2}{1-R^2}$$

3.2.2 Description of the Algorithm

The algorithm is described in the flowchart, Figure 3.1, which is essentially identical to Figure 2.2. The differences are in the definition of RSSQ, BRSSQ and the nature of the pivoting operation. Note that RSSQ and BRSSQ are standardized residual sums of squares.

1. MLEVEL is an array of p elements, each element corresponding to a variable. MLEVEL designates which variables are in the subset of N variables presently under consideration.

2. MBEST is an array similar to MLEVEL denoting which variables are in the best subset found so far.

3. NØIN is the number of variables presently pivoted in; NØIN < N at all times.

4. RSSQ and BRSSQ are defined on the variables not selected. Suppose N variables are presently under consideration, and all have been pivoted in. Let i be one of the (p - N) rejected variables, and r_{ii} be the i,i element of the transformed correlation matrix. Then,

$$RSSQ = \max_i (r_{ii}).$$

The minimum R^2 of the rejected variables regressed on a given set of N

Figure 3.1 - Flowchart for Interdependence Analysis

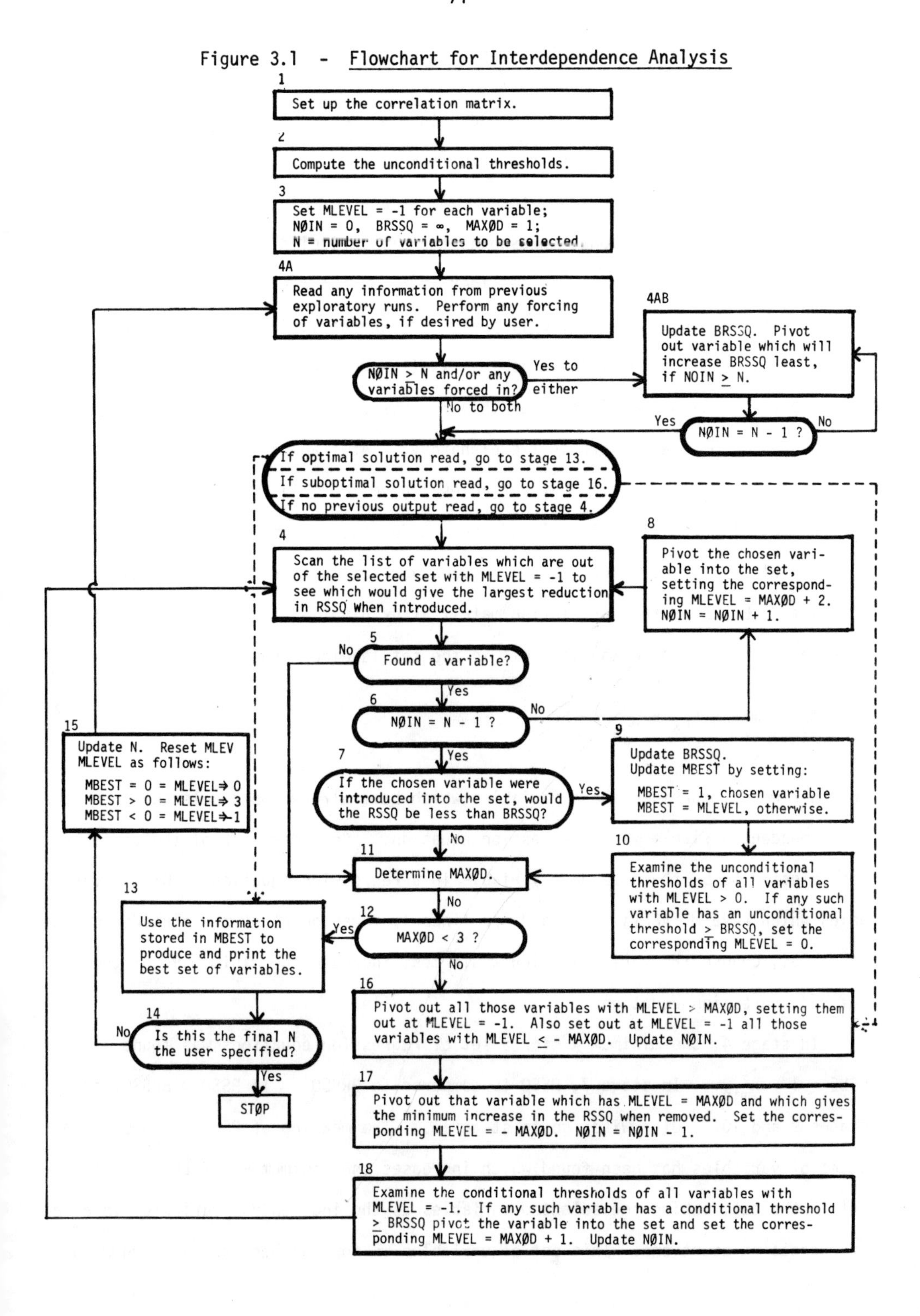

selected variables is 1 - RSSQ; i.e. RSSQ = $(1 - R^2_{min})$. BRSSQ is the same value for the best subset of N.

5. MAXØD is the maximum of the odd-valued elements of MLEVEL; if this results in a negative value, MAXØD = 1. When MAXØD = 1, the search for the best set has been halted and the result is printed.

In stage 1 of the algorithm, the p by p sample correlation matrix is computed. Stage 2 computes unconditional thresholds for all variables. To compute the unconditional threshold of variable i, consider it to be the dependent variable in a regression equation with all of the linearly independent variables except variable i. The unconditional threshold of variable i is $(1 - R^2)$ where R^2 is the square of the multiple correlation coefficient for this regression equation.

Stages 3, 5, 6 and 8 are self-explanatory. At stage 4, a variable is chosen by the following criterion. Let variables j have MLEVEL(j) = -1 and be linearly independent of those variables already chosen for the set under consideration. Consider all variables k such that $k \neq j$ and variable k is not already chosen for this set. Let R be the correlation matrix after pivoting in the variables already selected. Then, choose the variable in the set labeled j such that

$$\min_{j} \left[\max_{\substack{k \\ k \neq j}} \left(r_{kk} - \frac{r_{kj}^2}{r_{jj}} \right) \right]$$

In other words, consider the m variables which have already been selected to be the m independent variables in a regression equation. Consider each of the variables not selected to be the dependent variable in a regression equation. Of the variables which have MLEVEL(j) = -1 and are linearly independent of the already selected variables, choose the variable which would most increase the minimum R^2 of the defined regression equations.

In stage 4, the minimum R^2 for a set of regression equations was found. Let RSSQ = $1 - R^2_{min}$. In stage 7, RSSQ is compared to BRSSQ. If RSSQ < BRSSQ, do stages 9 and 10. Otherwise, go to stage 11. The meaning of RSSQ < BRSSQ is that a set of variables has been found which increases the minimum R^2 of the variables not in the set, taking the selected variables as the independent variables in a set of regression equations. This set of variables is now the best so far. On the other hand, RSSQ $\geq$ BRSSQ indicates the set of variables associated with (1 - BRSSQ) is better than the present set with R^2_{min} as calculated in stage 4.

In stage 9, BRSSQ is updated by setting BRSSQ = RSSQ. The purpose of stage 9 is to record in array MBEST the new set of variables which gives better results than any previous set. Stage 10 uses the unconditional threshold rule such that only a subset of the $\binom{p}{N}$ ways of choosing N variables to represent p variables need be examined. If k is one of the selected variables in the best set thus far, and if the unconditional threshold of variable k is greater than or equal to BRSSQ, then variable k must be included in every set chosen in the future. If these conditions are met, MLEVEL(k) is set to zero, which means that it is selected in every combination of variables in the future.

Stages 11, 12, 14 and 15 are self-explanatory. Stages 16, 17 and 18 are used to find a new set of N - 1 variables when the optimal solution is still uncertain. Stage 16 is easily understood except for one aspect; the variables with MLEVEL $\leq$ - MAXØD are not presently among the chosen variables. Setting their MLEVEL to minus one simply makes them available for selection in stage 18 or stage 4.

Stage 17 is relatively straightforward. Consider all of the N variables as independent variables in a regression equation, and each of the p - N rejected variables individually as a dependent variable in a regression equation. This gives p - N regression equations, each having an R^2. Let min R^2 be the minimum of the p - N values of R^2. The program chooses N variables such that min R^2 is greater than or equal to the min R^2 of any other set of N variables. Stage 17 chooses that variable with MLEVEL = MAXØD which gives the minimum increase in RSSQ when removed. That is, of the N - 1 selected variables, stage 17 chooses the one variable with MLEVEL = MAXØD which will least decrease the min R^2. This variable is the variable with MLEVEL = MAXØD which contributes least to the set of N - 1 variables.

Stage 18 computes conditional thresholds. To compute the conditional threshold of variable k with MLEVEL(k) = -1, form a set S of variables i, such that MLEVEL(i) = -1, or MLEVEL(i) < - MAXØD, or i is one of the currently selected variables, and such that the variables in S are linearly independent. Set Q is the set of variables j such that $j = k$, or j was requested by user to be a rejected variable, or $- MAXØD \leq MLEVEL(j) < - 1$. Let each of the variables in S, except for k itself, be independent variables in a set of regression equations on the elements of the set Q as dependent variables. Each of the defined regression equations has a R_j^2, the square of multiple correlation coefficient of the equation with variable j as dependent. The conditional threshold of variable k is maximum of $(1 - R_j^2)$, where j is an element of Q. Variables with conditional thresholds greater than or equal to BRSSQ must be in the selected set, as shown in stage 18 of the flowchart. The use of conditional thresholds in stage 18 is another device to reduce the total number of combinations of variables checked.

Stage 7, besides performing the functions already described, prints information to explain how the search for the best N variables of the p variables is proceeding. It prints the numbers of the N variables presently being considered and either that they will be introduced or not introduced. The message that they will be introduced means the combination just examined is the best thus far and that stages 9 and 10 will be executed. If they are to be introduced, the previous BRSSQ and min R^2 are printed as well as the new BRSSQ and new min R^2. The message that the combination of variables will not be introduced indicates that this set is not as good as a previous combination of N variables; therefore, the program proceeds to stage 11. If the variables will not be introduced, the program writes the min R^2 with this combination and $1 - R^2$, as well as the BRSSQ and the min R^2 of the best set of N variables thus far.

Stage 13 prints the final solution of N variables: value of N; numbers of the N variables selected; and min R^2 of this combination with the rejected variables. Also, the program computes which of the N variables taken as the dependent variable with the remaining N - 1 taken as independent, has the maximum R^2, as a measure of the extent of multicollinearity within the selected set. It prints this variable and the R^2 for the regression equation. As optional output, the program can print the p - N regression equations taking the selected variables as independent and the p - N rejected variables as dependent. The maximum likelihood estimate and corrected estimates of R^2, R, and F-ratio as well as the estimated standard error on R^2 are printed for each of the p - N regression equations. Another option simply lists the rejected variable, R^2, and F-ratio of regression equation, for each of the p - N regression equations.

In stage 13, the optimal solution must be prepared, since the last combination checked may not be optimal. Consequently, any variable selected in the final combination, but rejected in the optimal solution, must be pivoted out. Any variable rejected in final combination but selected in optimal solution must be pivoted in.

Stage 4A could be pictured as part of stage 4. Stage 4A is executed only the first time that stage 4 is entered immediately after N has been incremented. Stage 4A is used when results from a previous run are to be read in, or when the user has requested special treatment for some variables. This concept will be explained in Section 3.4. Essentially, this subroutine reads information from a previous run, makes all preparations to continue the search, and can force a variable to be included in the final selected set or excluded from the final selected set.

3.3 Example of Interdependence Analysis

Unlike the optimal regression program, no extensive experience with the interdependence analysis program can be reported. When the program was first written, a thorough comparison with principal components analysis was planned. However, this research was not undertaken. Therefore, the examples reported here are quite simple, and may not adequately display the potential, as yet untested, of interdependence analysis.

The 16 environmental chemical variables, which formed the set of independent variables in the case studies in Section 2.5, were subjected to interdependence analysis. The correlation matrix for the 16 variables was given in Table 2.5. Subsets with 1, 2, 3, 4, 6 and 8 variables were selected. The results, including the number of iterations required to find each optimal subset, are shown as Table 3.1. The total CPU time required to find all 6 optimal subsets was 35 seconds of IBM 360/75 time. Allowing 2 seconds to set up the problem and output the results, this amounts to 0.12 seconds per iteration.

Table 3.1

Example of Interdependence Analysis

	Number of Variables Selected											
	1		2		3		4		6		8	
R^2 and F with selected variables, S:	R^2	$F_{1,36}$	R^2	$F_{2,35}$	R^2	$F_{3,34}$	R^2	$F_{4,33}$	R^2	$F_{6,31}$	R^2	$F_{8,28}$
Variables												
1. Cadmium	0.130	5.36	S	-	S	-	S	-	0.630	8.79	0.622	5.97
2. Chromium	0.252	12.11	0.286	7.00	S	-	S	-	0.507	5.31	0.526	4.02
3. Copper	0.046	1.75*	0.252	5.90	0.254	3.86	0.333	4.12	S	-	S	-
4. Iron	0.018	0.67*	0.124	2.48*	0.256	3.89	0.286	3.30	S	-	S	-
5. Lead	0.057	2.18*	0.264	6.27	0.302	4.90	0.295	3.46	0.412	3.62	0.579	3.89
6. Manganese	0.178	7.79	0.215	4.80	0.342	5.90	0.291	3.39	0.533	5.91	0.512	3.81
7. Nickel	0.396	23.60	0.596	25.83	0.603	17.24	S	-	S	-	0.536	4.19
8. Tin	0.133	5.53	0.354	9.61	0.367	6.57	0.385	5.16	0.429	3.88	S	-
9. Titanium	0.037	1.39*	0.150	3.09*	0.351	6.14	0.338	4.22	S	-	S	-
10. Vanadium	0.415	25.55	S	-	S	-	0.660	16.04	0.680	10.99	0.563	4.67
11. Zinc	0.217	9.98	0.463	15.11	0.509	11.75	0.513	8.71	S	-	0.664	7.16
12. Nitrogen Dioxide	0.212	9.70	0.287	7.06	0.290	4.64	0.394	5.36	0.427	3.85	S	-
13. Sulfur Dioxide	0.362	20.40	0.411	12.19	0.415	8.04	0.335	4.15	0.436	4.00	S	-
14. Particulate Sulfate	S	-	0.476	15.88	0.518	12.16	0.476	7.50	0.501	5.18	0.517	3.89
15. Water Hardness	0.023	0.84*	0.202	4.44	0.204	2.91	S	-	S	-	S	-
16. Arsenic	0.144	6.06	0.254	5.96	0.261	4.01	0.313	3.76	0.412	3.62	S	-
Minimum R^2 with rejected variables: (variable, R^2)	4,	0.018	4,	0.124	15,	0.204	4,	0.286	16,	0.412	6,	0.512
Maximum R^2 among selected variables: (variable, R^2)	not applicable		10,	0.034	2,	0.286	2,	0.289	11,	0.369	9.	0.445
Number of iterations	1		11		38		71		67		79	

* not significant at α = 0.05 level of significance

As with regression analysis, the results show that different subsets of variables are selected as N is increased. For example, variable 14 chosen for N = 1, is not selected again. Variables 1 and 10 chosen for N = 2 are retained for N = 3 in a stepwise manner, but then discarded in N = 6 and 8.

The maximum R^2 among the selected variables appears to provide an indication of what selected variables may be rejected in a subsequent equation; see N = 2, 3, 4, and 6. The number of iterations required for these problems seems quite reasonable. However, as the number of variables increases, the computation time increases sharply, at least in proportion to p^2.

3.4 Suggestions for a Strategy for Using the Program

Although the algorithm uses bounding rules to limit its search, the number of combinations of variables it must check is sometimes large . In several tests, on the IBM 360/75, the program generally required 0.1 to 0.2 seconds per iteration. Therefore, the program can use large amounts of time when many of the combinations of selected variables are about equally good descriptors of the rejected variables. Two procedures are provided in the program to limit execution time.

First, the user can specify each value of N for which a set of variables is desired. For example, suppose there are 25 variables. By setting N = 5, 10, 15 on a preliminary run, the values of N which are of interest could be located. Suppose for N = 5, the min R^2 was extremely low; for N = 10, the min R^2 moderately small; while for N = 15, the min R^2 was very high. Then, the user may want to perform another run to find the results for N between 10 and 15.

Second, the user may limit the number of combinations of variables examined for a given N by setting an iteration limit. The program requires the user to specify such a maximum for each value of N desired. The rationale is that if the program is examining many combinations of variables, probably none of the combinations would give a min R^2 very different from the others because they probably have nearly equal descriptive power. Therefore, a limit on the length of the search may not seriously change the results, at least in terms of R^2.

However, there is a very serious problem with such a procedure. Suppose the user specifies three values of N: N_1, N_2, N_3 where $N_1 < N_2 < N_3$. The search for N_2 starts with the N_1 set of variables. If the search for N_1 variables was stopped prematurely, and if the search for N_2 variables is stopped before completion, the result for N_2 may be far from the best solution. The effect of successive suboptimal results may mount, causing the program to fail badly in selecting the best subset of N_3 variables. Furthermore, if both N_1 and N_2 yielded suboptimal results, the N_2 variables used to start the search for N_3 variables may be far from desirable.

To provide flexibility in dealing with this problem, the user can request punched output describing if the result is optimal or not, and the place where the search was stopped. If the search was stopped by the iteration limit, the search can be restarted in a subsequent run of the program precisely at the point where it was stopped. Also, optimal results can be read in from card output and the program will print the result without conducting the search. For explanation of the use of this capability, see Sections 3.5.2, 3.5.4, 3.5.10, and 3.5.11.

3.5 Order and Detailed Description of Input Card Types

The input cards are arranged in the following order:

1. Title
2. Problem definition
3. Values of N
4. Sets previously processed (optional)
5. Iteration limits
6. Variables to be forced (optional)
7. Format of data
8. Output options
9. Data deck
10. Unconditional thresholds (optional)
11. Information from previous jobs (optional)

The program can process several problems in one job: set up the first problem as shown, followed by the title card and card types 2-11, as needed, for the second problem. After finishing a problem, the program will try to read a new title card. Consequently, the program continues to read input for problems until there is no input. At this point the program reads an end of file in the card input and the computer may print an error message. However, this is the normal way the program stops and is no difficulty.

3.5.1 Title Card - FØRMAT(20A4)

Any 80 characters may be used to identify the problem.

3.5.2 Problem Definition - FØRMAT(5I5,F10.8,4I5)

NVAR - total number of variables.

NCASES - number of observations in the data deck.

NMAX - number of values of N to be specified; if NVAR is 25 variables, and subsets of 5, 10, and 15 variables are desired, NMAX = 3.

NFMT - number of cards required to specify format of data deck.

II - a binary indicator;

0, intermediate output is desired; intermediate output is for debugging purposes only.

1, normal option; no intermediate output.

EPSLØN - tolerance level; if several variables are nearly linearly dependent, the pivot operation would be undefined because of division by zero. Hence, EPSLØN is a tolerance level to prevent division by an amount less than EPSLON. 10^{-4} or 10^{-5} are recommended.

NKNØWN - indicates whether parts of the problem have been worked on in previous jobs, and is used chiefly if a search was stopped prematurely before finding optimal result. See also card types 4, 10, and 11, and Section 3.4. The punched output from the previous run is required as input to this problem if the search is to continue where it stopped.

0, normal.

i, where i > 0, indicates that information for i values of N, which were previously completed or partially completed, are to be read.

IFØR - an option for forcing variables into any set of N, or for excluding variables from any set N.

0, no forcing of variables.

i, where i > 0 indicates i variables are to be forced.

IPRINR - a ternary indicator.

0, print the minimum output, consisting of variables selected, min R^2 with rejected variables, max R^2 among the selected variables and the dependent variable for the max R^2.

1, print minimum output plus the following: R^2 using each rejected variable as dependent, standard error on R^2, F-ratio.

2, print minimum output plus the following: full regression equations using N selected variables as independent and each rejected variable as dependent and the output with IPRINR = 1 plus corrected estimate of R^2 and F-ratio for each equation.

IPUNCH - a binary indicator.

0, do not punch card output.

1, punch card output so that it may be used in subsequent runs with NKNØWN > 0.

3.5.3 Values of N - FØRMAT(16I5)

On this card, list in fields of I5 the NMAX values of N desired for this problem. The program will first find a subset of N_1 variables from the set of NVAR variables such that N_1 is the best subset according to the criterion. Then, it will select the best subset with N_2 variables. Specify the values of N in ascending order. If NMAX > 16, continue the values of N on a second card. Values are read into array LIST.

3.5.4 Sets Previously Processed - FØRMAT(16I5)

This card type should only be included if NKNOWN $\neq$ 0. This card type is used only when partially complete or complete optimal solutions are available from previous runs of the program. The values for this card are the values of N which were previously processed. The program will expect a card type 11 for each N specified. There must be NKNØWN of these values of N, where NKNØWN is defined on problem definition card. If NKNØWN > 16, continue the values on a second card. Specify the values in ascending order. These values are read into array IKNØWN.

For example, consider a problem with 15 variables. On an exploratory run, the program found the best subset with 5 variables but failed to find the best solution with 10 variables because of the limit on the number of combinations of variables tried. Punched output was requested for that job. Now, the user wants to find the best subsets with 1, 3, 4, 7, 8, 9, and 10 variables. To have all the results together, he wants to read the solution for N = 5 variables and have it printed. Thus, NVAR = 15, NMAX = 8, NKNØWN = 2, because N = 10 partially done before and N = 5 completed before. The values of N for card type 3 are 1, 3, 4, 5, 7, 8, 9, 10; the values for card type 4 are 5 and 10. The program will then expect card input punched by the program in a previous job for N = 5 and N = 10. See also Section 3.4, card types 10 and 11, and NKNØWN and IPUNCH of card type 2.

3.5.5 Iteration Limits - FØRMAT(16I5)

Iteration limits restrict the number of combinations of variables examined in the search; for each value of N, the program requires an iteration limit. If the number of iterations reaches the limit, the program immediately stops and prints the best set found so far. In the output of this set, the program will print "NØNØPTIMAL RESULT". This signifies that the program stopped the search before completion and that the result may be suboptimal. If the search had stopped normally before reaching the limit, the program prints "OPTIMAL RESULT" to indicate that the result is the best possible.

NMAX of these limits are required. The first value of first card corresponds to N_1, and the sixteenth on first card corresponds to N_{16}. If NMAX > 16, continue the iteration limits on a second card. All iteration limits should be greater than zero. Iteration limits are read into array NITL.

3.5.6 Variables to be Forced - FØRMAT(16I5)

Include only if IFØR > 0 on problem definition card. The program gives the user the option of choosing several variables to be forced into the selected subset of N variables, whether they would give the optimal result or not. Also, the user may specify several variables to be excluded, i.e., forced not to be among the N selected variables; however, in choosing the N variables, R^2 of the excluded variable is taken into account. The program then seeks the best solution, given that the user has specified variables forced to be in the selected subset or excluded from the selected subset.

The program provides the user with three options:

1. Force variable i to be selected for each N listed on card type 3;
2. Force variable i to be selected for certain values of N listed on card type 3;
3. Force variable to be excluded for certain values of N listed on card type 3.

Options 2 and 3 may be combined for the same variable. To implement 1, place the integer i in the first field of 5 columns, and 100 in the second field. To implement 2 and/or 3 for a given variable i and given N of card type 3, prepare one card: place i in the first field of 5 columns of the card; to force i into a selected subset of N, place an N in the next field; or to force variable i to be excluded from the subset of N, place -N in the next field.

Consider the following three examples and their interpretations:

column:	5	10	15	20
card 1:	4	-2	7	-5
card 2:	1	100		
card 3:	6	5		

1. Variable 4 is forced to be included in the subset of 7 variables, but will be excluded from subsets of 2 and 5 variables.
2. Variable 1 will be included in every subset: N = 2, 5, 7.
3. Variable 6 will be included in the subset of 5 variables.

For each variable to be forced, only one card of type 6 should be used. Since IFØR on card type 2 gives the total number of forced variables, there should be IFØR cards of type 6. The order in which the variables are specified does not matter, nor does the order in which the N's are specified.

Never force N variables into a subset of N; always specify less than N for a particular subset of N. Otherwise, the program prints the error message "SITUATIØN HØPELESS, BECAUSE CANNØT PIVØT ANY VARIABLES ØUT", the value of N and value of NØIN, the number of variables currently selected, and then the program stops.

Suppose in a previous job the user had IFØR > 0 and requested punched output. Then, for any subset where variables were forced in or out, those variables must again be specified on card type 6 when that card output is read as input. Thus, punching of card type 11 can be affected by variable forcing; however, punching of card type 10 is unaffected.

Values are read into array KLEVEL. The maximum number of variables that can be forced is 15; therefore, IFØR $\leq$ 15 and at most 15 cards of card type 6 may be included.

3.5.7 Format of Data - FØRMAT(20A4)

The data for the analysis can have almost any format, because the program requires the user to specify the format. The preparation of this card follows the rules for FORTRAN IV variable format cards. Consequently, the first character on first card should be a left parenthesis. The total length of the specification cannot exceed 400 characters (5 cards). If the expression is greater than 80 characters, continue it on a second card. The total number of cards to express the format must be given on the problem definition card under NFMT. The final character of the expression must be a right parenthesis. The expression is stored in array FMT1.

The data are read into a double precision array; therefore, only F and D type format specifications are allowed. If the data are punched as integers, they must be read as real numbers.

3.5.8 Output Options - FØRMAT(16I5)

NAPRNT - 0, do not print A-matrix.

1, print the A-matrix.

The A-matrix, proportional to the covariance matrix, is defined as follows for the i,jth element:

$$A_{ij} = \sum_{k=1}^{NCASES} (x_{ki} - \bar{x}_i)(x_{kj} - \bar{x}_j)$$

where $\bar{x}_i$ is the mean of the ith variable, x_{ki} is the kth observation on the ith variable, and NCASES is the number of observations.

NRPRNT - 0, do not print correlation matrix.

1, print correlation matrix.

NMEANP - 0, do not print means of variables.

1, print the mean of each variable.

NØP - 0, do not print observation matrix.

1, print observation matrix. This is relatively slow.

3.5.9 Data Deck - User Specified Format

The program reads the variables one observation at a time. The data must be organized such that the values of the variables for one observation are read in by one traversal of the variable format card.

Hence, the order would be:

$$\begin{matrix} x_{1,1} \cdots & x_{1,NVAR} \\ \vdots & \vdots \\ x_{NCASES,1} \cdots & x_{NCASES,NVAR} \end{matrix}$$

where NVAR is the total number of variables to be read, NCASES is the number of observations, and x_{ij} is the value of variable j for observation i. The values are stored in array X for only one observation. If the whole matrix need be stored, it is written on scratch tape, FORTRAN logical unit 9. Array X is double precision.

3.5.10 Unconditional Thresholds

These cards are punched during a previous job with IPUNCH = 1. They should be included in input only when NKNØWN ≠ 0.

1. The first card is a label for the convenience of the user. "THRESHOLDS -- I. A. VERSION 5" will be punched in columns 3 - 31.
2. The second card has the unconditional thresholds for variables 1 through 6. Third card has unconditional thresholds for variables 7 through 12; and so on. The format is (2X,6D13.5).

The unconditional thresholds are extremely important; by their use, the program examines only a subset of the total number of combinations of variables. If unconditional thresholds are read in by the program, they will not be punched again even if IPUNCH = 1.

3.5.11 Information from Previous Jobs

These cards were punched by a previous job with IPUNCH = 1. They are included as input only when NKNØWN ≠ 0.

This series of cards must be repeated NKNØWN times, resulting in NKNØWN sets. Each set was punched in the correct order by the computer, and corresponds to one of the values specified on card type 4, the sets previously processed card. The order of the sets must be the same as the order expressed on card type 4 and the N values must be in strictly increasing order.

A set of cards consists of the following:

1. Identifier - FØRMAT(3I5,D20.8)

NVAR - total number of variables.

N - number of variables desired for the set.

IØPT - a binary indicator.

0, search was halted because the maximum number of iterations was reached; results may not be optimal.

1, result is optimal, and the search was completed normally.

BRSSQ - value of BRSSQ when the search was stopped or ended.

2. MLEVEL - FØRMAT(16I5)

Punched only when IØPT = 0. The values of array MLEVEL are punched 16 per card, and indicate exactly where the search was ended when IØPT = 0.

3. MBEST - FØRMAT(16I5)

The values of array MBEST are always punched, 16 per card, and indicate the variables in the best set found so far.

This card type and card type 10 can be used for two purposes. If the iteration limits of card type 5 caused the search to stop before the program has determined the best subset of variables in a previous job, then these cards can be used to resume the search at the point where it was stopped. If the search ended normally with the optimal result having been found, the cards can be used to reprint the results; this enables results from several runs to be combined on one output for easy reference.

The cards are punched in correct order. A set is punched for each N in the job for which punching was requested. Therefore, one need only select the sets desired for input to the present job. If the value of N and NVAR on the identifier card do not correspond to the input values, the program stops and prints "ERROR -- DECK MIXED UP", along with the read values of N and NVAR.

3.5.12 WARNING!

There are no checks in the program to prevent errors in the input. For instance, the program would not balk if NVAR $\leq$ 0 or NCASES $\leq$ 0. However, this would have no meaning, and could cause the program to enter an infinite loop. Results may be unpredictable for incorrect input. Since there are no internal checks that parameters have meaningful values, the user should prepare the input carefully. The following limitations on variable ranges should be noted:

1. $0 < \text{NVAR} \leq 40$
2. $0 < \text{NCASES} \leq 1269$
3. $0 < \text{NMAX} \leq (\text{NVAR} - 1)$
4. $0 < \text{NFMT} \leq 5$
5. $0 \leq \text{NKNOWN} \leq \text{NMAX}$
6. 0 < variable format $\leq$ 400 characters
7. $0 \leq \text{IFØR} \leq 15$

3.6 Output of the Program

Initially, the program prints the values of user-specified input. A few of the output labels need to be explained. "NUMBER OF CASES" is the number of observations. The label for the values of N, the input from card type 3 is "CASES". This is followed by a list of the iteration limits and the problem title.

The user determines on the output options card whether the data deck, correlation matrix, means of variables, and A-matrix will be printed. If the data matrix need not be printed, the time to write and read observations from scratch tape and to print observations in output will be saved.

During the search for the optimal solution, the following output will be printed for each combination of variables considered:

1. The N independent variables under consideration.
2. If the set gives the best results yet, the program prints "WERE CONSIDERED AND INTRODUCED." The old BRSSQ with min R^2 and new BRSSQ with its min R^2 are printed. This indicates that the variables printed are the best solution found thus far.

3. If the set is not the best, the program prints "WERE CONSIDERED BUT NOT INTRODUCED" to indicate that the present set of variables is not as good as a previous set. The BRSSQ and min R^2 are printed for the best set. The min R^2 and RSSQ = $(1 - \min R^2)$ are printed for the set of variables just considered.

When the search is complete or when it is ended by reaching the iteration limit, the program prints the title of the problem, the value of N, and the variables selected. The min R^2 of these selected variables with the rejected variables is printed. Also, the variable in the selected subset with the largest R^2 with the remaining selected variables is printed with the value of max R^2. Additional output is available under options described in section 3.5.2.

In the output for each selected subset, a message is printed to indicate whether the search was completed or not. If the search ended normally and the result proved to be best, the program prints "ØPTIMAL RESULT". If the search was stopped because the iteration limit, the program prints "NØNØPTIMAL RESULT." When variables are forced as specified on card type 6, additional output is printed for each final result that includes forced variables. A list of variables forced to be excluded from the subset of N is printed as "SPECIFIED TØ BE IGNØRED." A list of variables forced to be in the subset of N is printed as "SPECIFIED TØ BE SELECTED." Also, a message "THEREFØRE, RESULTS MAY BE NØNØPTIMAL" is printed to indicate that better results may exist without having variables forced. These messages are independent of the messages "ØPTIMAL RESULT" and NØNØPTIMAL RESULT" for they tell whether the results are "optimal" or "nonoptimal" in the sense that the program completed its search or was halted by the user-specified iteration limit.

3.7 Machine Dependent Program Features and Suggestions for Modification

Sections 3.7 and 3.8 are concerned with programming technicalities, and will not usually need to be considered by users. The program was written in FORTRAN IV on an IBM 360/65. Unfortunately, the program does use some features which are unique to the 360, and, minor incompatibilities could arise if another computer is used. For instance, all variables declared as REAL*8 would have to be declared DOUBLE PRECISION; any variables declared REAL*4 should be declared as single precision REAL; variables declared INTEGER*4 must be declared INTEGER for other computers.

The READ statement at the beginning of the program has an END parameter. This READ inputs the title card. The END = n in the READ statement means that control will be transferred to statement number n if an end-of-life is encountered. For computers other than the 360, appropriate end-of-file procedures would have to be substituted. Other incompatibilities may exist; however, they would probably also be minor.

The program needs a scratch tape to be mounted on FORTRAN logical unit nine for storing in binary the data matrix when it is to be printed. The user should set up scratch tape on FORTRAN logical unit nine with appropriate control cards. The program writes one observation per logical record in binary on scratch tape for temporary storage. Consequently, the logical record length must be at least long enough to store 40 binary double precision variables. If the program should be expanded to permit NVAR > 40, the logical record must also be expanded. The maximum number of observations the program can handle is limited only by the amount of storage available on the scratch tape. There are no internal limitations on the number of observations. Consequently, the size of the scratch tape and the blocking set up for it are the only limitations on the number of observations (NCASES).

If more than five cards are needed to express the format of the observation deck, increase the dimension of the array FMT1 by 20 for each additional format card required. FMT1 appears in the main program and subroutine SETUP.

The maximum number of variables can be increased somewhat. To do so, compile the program, and note the amount of core used. Then, from the total amount of core storage available to users, find how much unused core remains. To increase the limit on NVAR to p, eight square matrices with dimension 40 by 40 must be increased to p x p. The arrays are RPRIME, A, R, ØRIG, STØRE, ATRIX, B, and C. RPRIME, A, and R are in labeled common and are declared in several subroutines. The other five arrays are local to one subroutine. ØRIG and STØRE are dimensioned in subroutine PIVØTR. ATRIX is dimensioned in RITØUT; and B and C are dimensioned in subroutine PLACE. Remember that each of the arrays is a double precision array, so each element in the array requires two locations instead of one. This procedure will permit a new limit p on NVAR such that NVAR $\leq$ p $\leq$ 75.

If there is still unused core and the problems require more than 75 variables, more revisions will be necessary. Suppose the new maximum for NVAR is to be p´, where p´ > 75. The eight square arrays mentioned previously must be increased to p´ by p´. All arrays which have dimension 75 or 80 must be increased to p´. Also, increase any vectors with 40 elements to p´ elements.

Increasing the limit of the number of variables which can be forced is not recommended. Not only would the array KLEVEL have to be increased to p by 18 for a new limit of p, but many of the DØ loop definitions would need a new upper limit of p instead of 15.

3.8 Description of the Program by Subroutine

The program consists of a MAIN program and 18 subroutines. Arrays and important variables not previously defined are documented in this section. Several of the subroutines correspond to the stages or boxes in Figure 3.1 in Section 3.2.2.

3.8.1 MAIN

MAIN controls the flow through the algorithm. Only the simplest stages are performed here; all other steps in the flowchart are done in subroutines. All input cards except card types 8 to 11 are read here. Important arrays are:

IKNØWN - values read from card type 4, values of N for which previous information is to be read from cards.

KLEVEL - columns 17 and 18 have a special purpose and are defined in other subroutines. For row i of columns 1-16;

if i $\leq$ IFØR, KLEVEL (I,J) corresponds to field j of card i of card type 6.

if i > IFØR, zero.

LIST - values of N specified by user read from card type 3.

NITL - iteration limits for each N read from input card type 5.

FMT1 - format of data specified by user read from card type 7.

TITLE - title as specified by user on card type 1.

Important variables are:

N4 - binary indicator used when NKNØWN $\neq$ 0;

0, information read in for this set of N variables may be suboptimal.

1, information from previous job is optimal for this N.

IREAD - binary indicator;

0, no information from early jobs for this reduced set of N.

1, information from earlier job was read for this set of N.

NITE - number of iterations performed for this value of N.

MØW - limit on number of iterations for this N.

BRSSQ - defined in Section 3.2.2.

IFOUND - variable selected at stage 4.

MAXØD - maximum positive odd value of elements in array MLEVEL; if negative, MAXØD = 1.

NØIN - number of variables pivoted in.

N - number of variables specified for selection.

MLIST - pointer to keep track of which element is to be used from arrays LIST and NITL.

The remaining variables were defined in the description of the input.

3.8.2 SETUP

This subroutine performs stage 1, makes almost all initializations, and reads card types 8 and 9, the output options card and data deck. It is the first subroutine called by the MAIN program. The data deck is read one observation at a time into array X. If the data is to be printed as a matrix, the program writes the observations directly on scratch tape, FORTRAN logical unit nine, and then processes each observation from scratch tape.

The means of variables are computed and stored in array XSUM. The correlation matrix is computed and stored in array R. In computing the correlation matrix, the sums of squares and cross-products matrix (A matrix) is also computed. As specified on output options card, the means, correlation matrix, data deck, and A matrix may be printed. Once the correlation matrix is computed, the program needs only the diagonal elements of the A matrix; these are stored in array ADIAG, and array A is freed to be a work array. Array R from this point on is always the correlation matrix with all variables i, given that MLEVEL(i) $\geq$ 0, pivoted in, except for those variables with MLEVEL = 0 and listed in column 17 of KLEVEL. If some of the independent variables are dummy variables, all column and row elements of correlation matrix are set to zero for the dummy variables, with the following exceptions: diagonal elements are set to one, and correlations between dummy variables are set to one.

3.8.3 STAGE2

If NKNØWN = 0, the subroutine computes the unconditional thresholds. If IPUNCH = 1, these are punched six per card; these cards could later be used for a job with NKNØWN > 0. To compute the unconditional thresholds, subroutine PRIMR is called. If NKNØWN > 0, the program reads unconditional thresholds from card type 10, and exits immediately.

Important arrays in this subroutine are:

T - unconditional thresholds.

R - original correlation matrix.

A - work array holding results of pivoting all linearly independent variables; initially used to compute the A matrix.

RPRIME - work array; if variable i is among the set of linearly independent variables, RPRIME is array A with variable i pivoted out.

ADIAG - diagonal elements of the original A matrix; contains sums of squared deviations about the variable means.

IMP - array containing the numbers of the variables which failed the tolerance (EPSLØN) test when computing array A. Those variables are linearly dependent with the variables pivoted into array A. For example, if variables 5 and 7 could not pass EPSLØN test, then IMP(1) = 5, IMP(2) = 7, and for all I $\neq$ 1 or 2, IMP(I) = 0.

3.8.4 PRIMR

PRIMR is called from STAGE2. Using the original correlation matrix with no variables yet pivoted in, the subroutine pivots in all linearly independent variables, storing the result in array A. Also, for any variable that could not be pivoted in because of linear dependencies, the variable's number is stored in array IMP. Array A is a work array here and not the A matrix in SETUP when computing the correlation matrix from the data. All other variables and arrays are as defined before. This subroutine simplifies computation of unconditional thresholds in stage 2.

3.8.5 STAGE4

Stage 4 is explained in the algorithm description, Section 3.2.2. Stage 4 selects the variable with MLEVEL = -1 that increases most the min R^2 of the selected variables with the rejected variables; the variable number is stored in IFØUND, and the associated min R^2 is stored by setting RSSQ = (1 - min R^2). The variable name for RSSQ is SMALL. R is the correlation matrix with all selected variables pivoted in. The rejected variables are those which either have MLEVEL < 0, or are listed in column 17 of KLEVEL with MLEVEL = 0. All other variables are selected variables.

3.8.6 CHKVAR

CHKVAR executes stages 7, 9 and 10, as explained in Section 3.2.2. Important arrays are:

- R - correlation matrix with all variables pivoted in with MLEVEL $\geq$ 0, but not listed in column 17 of KLEVEL.
- LVP - numbers of all selected variables except IFØUND; used to print the variables being considered for introduction as the best set of N, or rejection as suboptimal set.
- T - unconditional thresholds.
- MLEVEL - array of length NVAR giving position of the search and variable set being considered. If a variable i has MLEVEL(I) < 0, it is currently rejected, and not pivoted in. Variable IFØUND, however, does have MLEVEL = -1. If a variable i has MLEVEL(I) $\geq$ 0, it is currently selected and pivoted in, or user has forced it to be included or excluded for this N.
- KLEVEL - KLEVEL(I,17) lists variable user has specified to be excluded for this N; after the end of this list, entries are zero. KLEVEL(I,18) lists variables user has specified to be selected for this N, with zeroes after end of list. For columns 1-16, row i corresponds to card i of type 6, if i $\leq$ IFØR, and zero if i > IFØR.
- MBEST - array of length NVAR indicating which variables are in best solution so far. For variable i, if

  MBEST(I) > 0, then variable i is in best set so far;
  MBEST(I) < 0, then variable i is not in best set so far;
  MBEST(I) = 0, then variable i is a user-specified forced variable or has been definitely selected for best set by passing the unconditional threshold.

3.8.7 STAGE8

STAGE8 pivots in variable IFØUND and sets MLEVEL(IFØUND) equal to MAXØD + 2; NØIN = NØIN + 1. IFØUND and R are as defined in stage 4. All other variables are as defined before.

3.8.8 FMAXØD

This subroutine determines the value of MAXØD, the maximum positive odd MLEVEL.

3.8.9 ØUTAT1

ØUTAT1 performs stage 16 of the flowchart. All variables and arrays are as defined previously.

3.8.10 ØUTATM

This subroutine performs stage 17, which is explained in the description of the algorithm, Section 3.2.2. The procedure for finding the variable with MLEVEL = MAXØD which contributes least to the N-1 selected variables is given here. Let k be any variable with MLEVEL(K) = MAXØD. Let ℓ be a rejected variable. LSSQ = RSSQ is

$$\text{LSSQ} = \min_{k}\left\{\max_{\ell}\left(r(\ell,\ell) - \frac{r(\ell,k)^2}{r(k,k)};\ \frac{-1.0}{r(k,k)}\right)\right\}$$

The variable k which yields this value of LSSQ is the variable which is to be pivoted out. Thus, k is the variable which gives the minimum decrease in R^2_{min} of rejected variables when k is pivoted out, since R^2_{min} = 1 - LSSQ. That is, the formula maximizes the R^2_{min} assuming that one of the selected variables must be rejected.

3.8.11 CØNDTH

This subroutine performs stage 18 which computes the conditional thresholds defined in Section 3.2.2. If in computing conditional thresholds some variables cannot be pivoted because of linear dependencies, the numbers of these variables are stored in array IMP just as in STAGE2.

Conditional thresholds are assigned for each variable; however, if MLEVEL(K) $\neq$ -1, the conditional threshold of variable k is zero. The conditional threshold of the first variable with MLEVEL = -1 is computed. If the conditional threshold is greater than or equal to BRSSQ, the variable is pivoted in and its MLEVEL set equal to MAXØD + 1. If, as a result, NØIN = N - 1, the subroutine exits; however, if

NOIN < N - 1, the subroutine computes the conditional threshold of the next variable with MLEVEL = -1. It is therefore possible to exit the subroutine before computing all conditional thresholds. Any variables for which a conditional threshold is not computed are assigned a conditional threshold of zero. If NOIN < N - 1 even after all conditional thresholds have been computed, the program simply exits.

Computation of the conditional thresholds proceeds as follows. All selected variables plus all variables i with MLEVEL(I) < -MAXOD, or MLEVEL(I) = -1, are pivoted in leaving the result in array A. Any variables failing the tolerance test for division by zero are listed in array IMP. Suppose the variable whose threshold is being computed passed the tolerance test. Call this variable ℓ. Pivot ℓ out of matrix A, leaving the result in RPRIME. The conditional threshold is

$$\max_{M} \left(\text{RPRIME}(M,M)\right)$$

where M is any variable such that -MAXOD ≤ MLEVEL(M) < -1, or M was specified by user as excluded for this N. Recall the variables to be excluded for this N are listed in column 17 of KLEVEL and have MLEVEL = 0; not all variables with MLEVEL = 0 are variables user has forced to be rejected. For a variable which failed tolerance test, the same computation holds except that correlation matrix A is used rather than correlation matrix, RPRIME.

Important arrays are:

A - correlation matrix after all linearly independent variables i, given that (a)i is a selected variable, or (b) MLEVEL(I) < -MAXOD, or (c) MLEVEL(I) = -1 have been pivoted in.

R - correlation matrix with all selected variables pivoted in.

RPRIME - work array; when computing conditional threshold of a variable i that had been pivoted in to create array A, RPRIME is the result of pivoting out variable i.

CT - conditional thresholds.

For definitions of MLEVEL and KLEVEL, see Section 3.8.6. All other arrays and variables are as defined earlier.

3.8.12 PLACE (B,C)

The subroutine transfers the elements of array B to array C. As used in this program, R, A, and RPRIME are the only arrays that can be used as arguments in a call to PLACE.

3.8.13 PIVØTR (ØRIG, STØRE, NG, INVRNØ, IDIM, IDØNE)

This subroutine performs all pivot operations. The arguments are:

ØRIG - input matrix for the pivot operation.

STØRE - return matrix which will contain the result of the pivot operation; STØRE and ØRIG may be the same array.

NG - variable to be pivoted in or out.

INVRNØ - 0, means pivot variable NG in.

1, means pivot variable NG out.

IDIM - dimension of square matrices ØRIG and STØRE; this will be NVAR in each call of PIVØTR.

IDØNE - return variable;

if 1, pivot operation was not performed because of EPSLØN.

if 0, pivot was performed.

The EPSLØN test is only used to check variables that are to be pivoted in, as a variable already pivoted in can always be pivoted out.

Array SAVE is used internally in the program to store column NG of array ØRIG. Use of SAVE enables ØRIG and STØRE to be the same matrix in any call to PIVØTR.

3.8.14 RITØUT (ATRIX, INØUT, NEL)

This subroutine prints the result of pivot operations. Since this produces extensive output, it is only to be used when debugging the program. It is always called, but it exits immediately if II = 1. If II = 0, the subroutine prints the contents of a square matrix. The arguments are:

ATRIX - matrix to be printed

INØUT - 0, prints ATRIX and that variable NEL was pivoted in;

1, prints ATRIX and that variable NEL was pivoted out;

>1, prints ATRIX, but no messages;

NEL - variable that was pivoted in or out.

3.8.15 RESET

RESET performs stage 15 of the flowchart. However, the next value of N is assigned in the main program.

3.8.16 ØUTPUT

The subroutine executes stage 13, as described in Section 3.2.2. One additional point is now explained. If NKNØWN ≠ 0, that is if there is information from previous runs, ØUTPUT will be called from STG4A if the set of N variables read is not known to be optimal. When this happens, N4 = 0. The program automatically sets N4 = 0 in these cases, and does the following:

1. Pivots in variables i with MLEVEL(I) < 0 and MBEST(I) ≥ 0.
2. Pivots out variables j with MLEVEL(J) ≥ 0 and MBEST(J) < 0.

The purpose is simply to prepare the program to resume the search at the point where it was stopped in the previous job. The subroutine will exit without printing. N4 = 1 indicates that the program should print. Also, concerning the above pivot operations with respect to variable i:

1. ØUTPUT assumes that user-forced variables do not then have MLEVEL = 0 nor MBEST = 0.
2. MLEVEL is not updated as these pivot operations are made. Consequently, after all pivot operations are complete, MBEST not MLEVEL, corresponds to the variables now pivoted in.
3. If ØUTPUT was called from MAIN program, subroutine RESET (stage 15) will update MLEVEL to correspond to the variables currently selected.
4. If ØUTPUT was called from STG4A, then subroutine STG4A will update it.

When ØUTPUT is called from MAIN, ØUTPUT prints the user-forced variables for this N, if there are any. Remember that column 17 of KLEVEL lists variables forced to be excluded for this N, with zeroes following end of list. Column 18 of KLEVEL lists variables forced to be selected for this N, with zeroes following end of list.

Important arrays are:

- R - correlation matrix with all selected variables pivoted in.
- TITLE - user specified title
- COEF - array containing the coefficients of the variables, if regression equations are to be printed.

Ø - unused

RSD - unused

P - unused

IA - unused

Important new variables are these:

RS - R^2_{min} = 1 - BRSSQ; minimum R^2 with rejected variables.

RSL - maximum R^2 among selected variables.

BETA0 - constant of regression equation, if regression equations are computed.

BETAS - temporary storage for one coefficient, if regression equations are computed.

STDERR - used in computing standard errors of several statistics.

AR - square of multiple correlation coefficient of a regression equation; used when IPRINR = 1 or 2.

ARS - R, ARS = $(AR)^{\frac{1}{2}}$

F - used to compute both the maximum likelihood and corrected estimate of F-ratio, if both are computed.

BR - corrected estimate of square of multiple correlation coefficient; computed when IPRINR = 2.

BRS - corrected estimate of R, BRS = $(BR)^{\frac{1}{2}}$

BRSSR - residual sum of squares for a regression equation.

DMF - floating point variable, which is NCASES-N-1.

3.8.17 STG13A

This subroutine performs part of stage 13 and is called from the MAIN program immediately before calling subroutine ØUTPUT. One purpose of this subroutine is to prepare the program for the next N, if there had been forced variables for the current N, as follows:

1. setting MLEVEL = MBEST = -1 for any variables forced to be rejected for current N.

2. setting MLEVEL = MBEST = 3 for any variables forced to be selected for current N.

A second purpose of this subroutine is to give punched output if IPUNCH = 1. If IPUNCH = 0, the subroutine simply exits. When IPUNCH = 1, the program punches output which can be read in a later job. The subroutine punches card type 11 for each N processed. The output is described in detail in Section 3.5.11, which discusses using the cards as input to another job.

The following variables are important:

NITE - number of iterations examined for this value of N.

MØW - iteration limit for this N. If NITE = MØW, the search was stopped because the iteration limit specified by user on card type 5 was reached. Therefore, solution may be nonoptimal.

IFØR - number of variables to be forced.

3.8.18 STG4A

This subroutine is a preliminary stage to set up the problem just before entering the iteration loop. Important arrays are:

R - correlation matrix with all selected variables pivoted in.

KLEVEL - information concerning forced variables.

Important variables are:

IREAD - binary indicator:

0, no information was read

1, information was read

N4 - binary indicator: in the punched output, a variable IØPT was punched by stage 13 of an earlier job; N4 is the location into which IØPT is read.

0, search should be continued on this set; solution may not be optimal.

1, search was completed on this set; the solution is optimal.

Recall the following definitions given earlier:

NKNØWN - variable which user specifies on card type 2; NKNØWN is the number of sets for which there is card input from a previous job.

IKNØWN - array whose values are given by the user on card type 4; these are the values of N for which previous information is to be read.

IFØR - given by user on problem definition card.

If NKNØWN = 0 as specified by user, the subroutine will skip the coding dealing with reading previous results. N4 = 1 and IREAD = 0 at the time of exit. If NKNØWN ≠ 0, the subroutine searches through IKNØWN to find if this N is one of the sets for which previous information is available. If it is not, the subroutine exits with N4 = 1 and IREAD = 0. If this N is one of the IKNØWN values, card type 11 is read. If the punched values of NVAR and N do not correspond to the actual values the program expects, it stops after printing the value of NVAR (labeled N1), the value of N (labeled N3) as read from card type 11, and a message that input is mixed up.

If N4 = 0, as read from cards, i.e. if the search should be started at point where stopped, the MLEVEL cards are read into array MBEST. Then, ØUTPUT is called to make use of ØUTPUT's two procedures:

1. if MLEVEL(I) $\geq$ 0 and MBEST(I) < 0, pivot out variable i; pivots out currently selected variables which should be rejected.
2. if MLEVEL(I) < 0 and MBEST(I) $\geq$ 0, pivot in variable i; pivots in currently rejected variables which should be selected.

Since the MLEVEL's are those from the last value of N, and since the MBEST's are the new values of MLEVEL, calling ØUTPUT has the effect of reselecting those variables that were selected when the search was stopped. ØUTPUT will do no printing since N4 = 0. Then, the MLEVEL array is given the values of MBEST, so the MLEVEL array read into MBEST is finally put into MLEVEL. The rest of the subroutine is the same as if N4 = 1. When the subroutine exits, N4 = 0; IREAD = 1.

The remainder of the subroutine is executed whether N4 = 0 or 1. The MBEST cards are read into array MBEST. If the solution read was optimal, the subroutine has N4 = 1 and IREAD = 1. If card type 11 is read, MAXØD = 1 whether N4 = 1 or 0.

Next there is a block of coding dealing with user-specified forced variables. If IFØR = 0, the coding is skipped, for there are no forced variables for these observations. This block of coding basically performs four operations:

1. finds all variables forced at this N, setting their MLEVEL = MBEST = 0;
2. for all variables specified by user as forced to be excluded for this N, lists each variable in column 17 of KLEVEL, making sure zeroes follow the end of the list, and pivots the variable out if it is currently in;
3. for all variables specified by user as forced to be selected for this N, lists each variable in column 18 of KLEVEL, making sure zeroes follow the end of the list, and pivots the variable in if it is currently out;
4. executes subroutine STG4AB if necessary.

If the user specified a variable to be selected, but it fails the linear independence test, a message will be printed: "CØULD NØT PIVØT IN VARIABLE", variable number, "IN STG4A". The program will continue but not force the variable to be selected.

Note that all entries after the end of the list in column 17 of matrix KLEVEL must be zero, and all entries after the end of list in column 18 of matrix KLEVEL must be zero. Whether there are any variables forced or not, whether IFØR = 0 or not, this must hold because code in other subroutines checks columns 17 and 18 of KLEVEL, expecting zeroes after the end of each list. STG13A is the complement of STG4A; it brings forced variable manipulations back to normal and punches card type 11.

3.8.19 STG4AB

This subroutine is called only from subroutine STG4A. If subroutine STG4A read no card type 11 for this N, and if the number of selected variables $\geq$ N, or there are forced variables for this N, then STG4A calls STG4AB to perform three operations:

1. update BRSSQ;
2. remove nonforced selected variables until number of selected variables < N;
3. update MBEST to correspond to BRSSQ.

When variables are forced, the best residual sum of squares for the previous N could be better than any subset with N variables when there is forcing. Also, with variables already selected for the previous N, forcing in variables for the current N could cause the number of selected variables to be $\geq$ N; therefore, STG4AB is called.

BRSSQ is updated first. If NØIN < N, the subroutine performs the third operation above; otherwise it tries to remove one of the nonforced selected variables. If there are none which can be removed, an error message is printed, and the program stops. If a variable can be removed, it is removed and its MLEVEL set equal to MBEST = -1. Computation of the variable to be removed is similar to that in ØUTATM.

Before returning, STG4AB checks all nonforced variables with MLEVEL = 0, as part of operation 3. Since BRSSQ is updated and may have increased, the program must make sure that all nonforced variables with MLEVEL = 0 still have their unconditional threshold $\geq$ BRSSQ. If any such variable has its unconditional threshold < BRSSQ, its MLEVEL and MBEST must be set to 3.

4. OPTIMAL NETWORK ANALYSIS

4.1 Introduction

Following implementation of optimal regression and interdependence analysis, the authors noted that the general algorithm could also be used to solve the optimal network problem which may be stated simply as follows: given n transportation nodes (cities, stations, or points to be served) and the set of n(n - 1)/2 direct, non-directional links joining all pairs of nodes, select the subset of links that minimizes the sum of shortest path distances between all pairs of nodes, subject to a budget constraint on the total link length. This application extended the original algorithm of Beale et al. (1967) in two ways. First, the size constraint (number of variables) is generalized to a linear constraint (total length of network). Second, the algorithm is applied to a potentially larger tree-search problem outside the field of statistics; however, Beale (1970) also described the application of the algorithm to a plant location problem.

In the optimal network problem, the transportation link directly connecting a pair of nodes replaces the variable as the element to be selected. The speed of the algorithm therefore depends on the ability to update the matrix of node to node distances (or travel times) following the addition or deletion of a link. Murchland (1967, 1970) had devised a highly efficient algorithm for this purpose which was readily incorporated as the "pivot" procedure. A minimum spanning tree algorithm by Sollin, described by Berge and Ghouila-Houri (1965), was also usefully employed for finding an initial solution.

As a detailed description of the generalization of Beale's algorithm to a linear constraint has been published elsewhere (Boyce, Farhi and Weischedel, 1973), this chapter is mainly a technical description of the algorithm and computer program. Some additional details of examples are also provided.

4.2 Optimal Network Algorithm

As the algorithm makes extensive use of procedures by Murchland and Sollin, these are described first, followed by the description of the optimal network algorithm.

4.2.1 Minimum Path and Minimum Spanning Tree Algorithms

Three related minimum path algorithms are incorporated into the optimal network program. The first finds minimum paths between all pairs of nodes by the Floyd-Murchland fixed matrix method. It is used only to compute the matrix of shortest distances for the initial solution; thereafter, the link addition and link deletion algorithms of Murchland are employed. The fixed matrix method is very convenient for the relatively small networks studied here. However, it is inappropriate for large networks with several hundred nodes; for such networks, tree-tracing methods are inherently superior. This section concludes with a description of the minimum spanning tree algorithm due to Sollin.

Minimum Paths Between All Pairs of Nodes

Consider a network with nodes $i,j = 1,\dots n$, and direct, directional links with positive length between certain nodes i and j; if link (i,j) does not exist, then its length is considered to be infinite. Find the matrix of shortest path distances between all pairs of nodes.

According to Dreyfus (1969), the methods of Floyd (1962), later rediscovered by Murchland (1965, 1970), and Dantzig (1966) are the most efficient known, assuming that the time to perform an addition is less than or equal to twice the time to perform a comparison. Both algorithms require $n(n-1)(n-2)$ additions and comparisons. The Floyd-Murchland method is very simple to program and takes advantage of the condition that for some links $d(i,j) = +\infty$.

To find the minimum distance between every pair of nodes (i,k) of an n node network, n iterations are required. Let

1. $d_0(i,k)$ = direct link distance from i to k;
2. For all j, $1 \le j \le n$, $d_j(i,k) = \text{minimum} \left[d_{j-1}(i,k),\ d_{j-1}(i,j) + d_{j-1}(j,k) \right]$.

$d_j(i,k)$ gives the minimum distance using only nodes m, $1 \le m \le j$, as intermediate nodes in a path between node i and node k. $d_n(i,k)$ is the final minimum distance between node i and node k.

The procedure, as coded, takes advantage of the restriction that the program is designed for symmetric networks. Also, the Floyd-Murchland procedure implemented computes for all node pairs (i,k), the first back node of a minimum path between nodes i and k. (The back node matrix is not used in the search for the optimal network, but may be useful for debugging or checking results.)

Initially, define two matrices, a distance matrix, F, and a back node matrix, B. The variable names used here are the actual arrays in the program. Let

$F(i,j) = \infty$, if $i \neq j$ and if link (i,j) is not included in the network;

0, if $i = j$;

distance or length of link (i,j), if $i \neq j$ and if link (i,j) is in the network.

$B(i,j) = i$ for all (i,j).

The procedure consists of three loops:

For all $j = 1 \ldots n$, perform the following:

for all $i = 1 \ldots n-1$ such that $F(i,j) < \infty$, perform the following:

for all $k = i+1 \ldots n$ perform the following:

if $F(i,j) + F(j,k) \geq F(i,k)$, do nothing:

otherwise, perform the four operations:

1. set $F(i,k) = F(i,j) + F(j,k)$,
2. set $F(k,i) = F(i,j) + F(j,k)$,
3. set $B(i,k) = B(j,k)$, and
4. set $B(k,i) = B(j,i)$.

Compare this procedure to the description with equations 1. and 2. above. In the procedure implemented, j plays the role of possible intermediate nodes used in the path from i to k. For each j, the comparison $F(i,j) + F(j,k)$ to $F(i,k)$ and the operations depending on the result of that comparison perform equation 2. At the completion of the procedure,

$$F(i,k) = \begin{cases} \text{minimum distance from } i \text{ to } k, & \text{if } i \neq k; \\ 0, & \text{if } i = k. \end{cases}$$

$$B(i,k) = \begin{cases} i, & \text{if } i = k \\ \text{first back node on a minimum path between } i \text{ and } k, & \text{if } i \neq k. \end{cases}$$

Minimum Path Distances After Deleting a Link

Consider a given subset of symmetric links defining a network. Suppose that the matrix F of minimum internodal path distances is known. A single link is deleted from the network; find the new minimum path distances.

Clearly, this problem could be solved using the above method by recomputing the minimum distances. However, two procedures by Murchland (1967, 1970) and Halder (1970) solve the problem more efficiently by using the additional information contained in matrix F.

Deciding which procedure is more efficient is difficult because the number of computations for both depends upon the particular network and upon the particular link chosen to be removed. Halder's procedure could have an advantage in efficiency over Murchland's method, particularly for the kinds of networks considered by the optimal network program. However, the existence of such an advantage seems difficult to determine theoretically, and may not be significant. Considerable computational experience on various kinds of networks seems necessary to compare the two methods. The procedure actually implemented differs slightly from the method Murchland proposed.

As above, let n be the number of nodes of the network, and DP the matrix of the former minimum path distances, which has zero diagonal. A second matrix PP has diagonal PP(i,i) = i; PP(i,j), $i \neq j$, is the first back node on a minimum path from i to j. Suppose link (jp, kp) is to be removed, and ℓ is its link length. Then the method implemented, which is modified from Murchland's method, is the following:

1. If DP(jp,kp) < ℓ, then stop, for the link could not have been used for any minimum path. (Note: Murchland (1967) has DP(jp,kp) $\leq \ell$, which is incorrect, for the minimum path may be link (jp,kp) if DP(jp,kp) = ℓ.)
2. For each node i,
   a) if DP(i,kp) = DP(i,jp) + DP(jp,kp), place node i in set M.
   b) if DP(jp,i) = DP(jp,kp) + DP(kp,i), place node i in set N.

Note: jp ε M and kp ε N always by above procedure. $N \cap M$ is empty if and only if $\ell > 0$. Also, if $\ell = 0$, every vertex satisfies both a and b. The coding of the algorithm takes advantage of these properties.

3. If link (i,m) is an element of the cross-product MxN, reset DP and PP in the following way:

   DP(i,m) = ∞, if the link is not included in the network without link (jp,kp);

   link length of (i,m), if the link is included in the network without (jp,kp).

   DP(m,i) = DP(i,m) by definition just given.

   PP(i,m) = i

   PP(m,i) = m

   This is done for all (i,m) ε MxN.

4. For each v = 1 ... n, and for each i ε M where $i \neq v$, and for each m ε N do the following:

   a) if DP(i,m) $\leq$ DP(i,v) + DP(v,m), do nothing.

   b) if DP(i,m) > DP(i,v) + DP(v,m), set

      1) DP(i,m) = DP(i,v) + DP(v,m)

      2) DP(m,i) = DP(i,v) + DP(v,m)

      3) PP(i,m) = PP(v,m)

      4) PP(m,i) = PP(v,i)

Step 1 is a check to see whether any of the minimum distances could be affected. Step 2 finds the set of all node pairs which could have used link (jp,kp) in the minimum path between the two nodes; that set of all node pairs is MxN. Step 3 initializes matrix DP. If (i,m) ε MxN, minimum path distance must be recomputed, for it may increase upon removal of (jp,kp). A first estimate of the new minimum distance is the updated value DP(i,m). This estimate is ∞ if link (i,m) is not in the constructed network, but the length of (i,m) if the link is in the constructed network. For any node pair not in MxN, the minimum path is not affected, for the minimum path could not have used link (jp,kp) by the way sets M and N are defined. Step 4 is the Floyd-Murchland method for computing minimum distances between all nodes, and is similar to procedure outlined above. The only difference is that minimum paths are only recomputed for node pairs in MxN, as all other minimum paths could not have changed. Matrix DP has the new minimum path distances of the network minus link (jp,kp). Matrix PP contains the first back nodes for the new minimum distances.

As a justification for this process, note that if a path between nodes j and k used link (jp,kp), then at least one of the following holds,

1. DP(j,k) = DP(j,jp) + DP(jp,kp) + DP(kp,k)
2. DP(k,j) = DP(k,jp) + DP(jp,kp) + DP(kp,j)

where DP is the matrix of old minimum distances. Hence, the set of all such node pairs is contained in MxN.

Minimum Path Distances after Adding a Link

Consider a given set of symmetric links defining a network. Suppose that the matrix F of minimum internodal distances is known for that network. A single link is added to the network. Find the new minimum internodal distances.

This problem could also be solved by the fixed matrix procedure. However, since the information concerning the former minimum distances is available, the problem can be solved much more efficiently.

The algorithm by Murchland (1967) was also implemented to solve this problem. Suppose link (j,k) is to be added to the network. Let D represent the matrix of minimum internodal distances with diagonal elements equal to zero. Let P represent the matrix of first back nodes corresponding to D, with diagonal elements P(i,i) = i. As above, the link length of link (j,k) is ℓ and the total number of nodes is n.

1. If D(j,k) $\leq \ell$, then stop, for adding link (j,k) cannot change any minimum distance, Otherwise, set
   a) D(j,k) = ℓ
   b) D(k,j) = ℓ
   c) P(j,k) = j
   d) P(k,j) = k.

2. For all i, $i = 1, 2 \ldots n$, if $D(i,k) > D(i,j) + \ell$, set

   a) $D(i,k) = D(i,j) + \ell$

   b) $D(k,i) = D(i,j) + \ell$

   c) $P(i,k) = j$

   d) $P(k,i) = P(j,i)$

   e) $i \varepsilon M$.

3. For all m, $m = 1, 2 \ldots n$, if $D(j,m) > \ell + D(k,m)$, set

   a) $D(j,m) = \ell + D(k,m)$

   b) $D(m,j) = \ell + D(k,m)$

   c) $P(j,m) = P(k,m)$

   d) $P(m,j) = k$

   e) $m \varepsilon N$.

4. For all $(i,m) \varepsilon MxN$, if $D(i,m) > D(i,k) + D(k,m)$, set

   a) $D(i,m) = D(i,k) + D(k,m)$

   b) $D(m,i) = D(i,k) + D(k,m)$

   c) $P(i,m) = P(k,m)$

   d) $P(m,i) = P(k,i)$.

Essentially, when a link (j,k) is added to the network, the minimum paths either remain the same or use link (j,k). Step 1 checks whether the link would be used at all, for if the minimum path between j and k is already shorter than the link length, no minimum path would use the link. Step 2 finds all nodes i which would use link (j,k) in a minimum path to node k. Step 3 finds all nodes m which would use link (j,k) in a minimum path from node j. Step 4 updates the paths between those pairs of nodes whose paths are changed in Steps 2 and 3.

Minimum Spanning Tree

The minimum spanning tree defined on the length or construction costs of links is used as an initial solution. An algorithm attributed to Sollin described by Berge and Ghouila-Houri (1965) was chosen because of its simplicity.

Given n nodes and $n(n-1)/2$ links joining them, let there be a cost, $c(i,j) \geq 0$, for including each link in the network. Link costs exist for every link, but may be infinite; however, every finite link cost must be unique. Let there be at least one spanning tree (a network of exactly (n-1) links connecting every node) such that the tree has finite total cost. Find the minimum spanning tree, i.e. the minimum total cost tree that connects every node.

Sollin's algorithm has four steps:

1. For every node i, $1 \leq i \leq n$, find the node j such that

$$c(i,j) = \underset{\substack{1 \leq k \leq n \\ k \neq i}}{\text{minimum}} \left\{ c(i,k) \right\}$$

Select the link (i,j).

2. Step 1 has created subnetworks or components of connected nodes. If there is only one such component, stop. Otherwise proceed to step 3.

3. Number the components, of which there must be ℓ, $2 \leq \ell < n$. Let A_i denote the set of nodes in component i, $1 \leq i \leq \ell$. For each component i, find the component j, $j \neq i$, $1 \leq j \leq \ell$ with nodes $m \in A_i$ and $n \in A_j$, such that

$$c(m,n) = \underset{\substack{1 \leq k \leq \ell \\ k \neq i}}{\text{minimum}} \left\{ \underset{\substack{m' \in A_i \\ n' \in A_k}}{\text{minimum}} \left(c(m',n') \right) \right\}$$

That is, find the component j which can be connected to component i by the least expensive link (m,n). Select link (m,n).

4. Form new components by the ℓ links found in step 3 (at least one is selected twice). If as a result, there is now only one component, stop. Otherwise, proceed to step 3 again with these new, larger components.

4.2.2 Description of the Algorithm

The algorithm is described in Figure 4.1, which is a specific application of the general algorithm given in Figure 2.1. At several places in the flowchart (stages 4, 8, 16, 17 and 18) a single link is added to or deleted from an existing network using the procedures described in Section 4.2.1. The minimum spanning tree is applied in stage 1.

Bounding Rules

Since the addition of a link causes the objective function to decrease or at worst remain the same, the two bounding rules are again useful in reducing the number of networks that need be examined. The objective function for the network problem is the sum of minimum internodal distances.

The unconditional threshold of a link (i,j) is the sum of minimum internodal distances of the entire network (every link constructed) except for link (i,j). If the objective function for the best network found so far is less than or equal to the unconditional threshold of link (i,j), then link (i,j) must be in the optimal network; otherwise, even with the entire network less link (i,j), the objective function is not better than the network already examined. Therefore, all networks without link (i,j) need not be examined. Stage 10 of the flowchart designates such a link (i,j) by setting its MLEVEL = 0.

Suppose a given network is being examined; to proceed through particular branches of the combinatorial tree, the conditional threshold may be useful for excluding a set of links S. To find the conditional threshold of link (i,j), compute the objective function of the entire network without link (i,j) and without the set of links S. If the objective function for the best network so far is less than or equal to the conditional threshold of link (i,j), then link (i,j) must be in the best network which excludes S. For without link (i,j), a better objective function cannot be found. Therefore, in examining networks without S, only networks containing link (i,j) need be examined. This feature is implemented in stage 18 of the flowchart. If the budget constraint allows a link (i,j) to be included, it is designated by setting its MLEVEL = MAXØD + 1, i.e. its MLEVEL is an even number.

Consider a network such that one or more single links could be added within the budget constraint, but not more than a single link can be added. Then, the effect on the objective function of adding a single link can be examined, and the link that most decreases the sum of minimum internodal distances selected. This link is known to be the best with respect to the previous network. At stages 7 and 9 this result is used in that the last link selected by stage 4 retains MLEVEL = -1.

Figure 4.1

Flowchart for Optimal Network Analysis

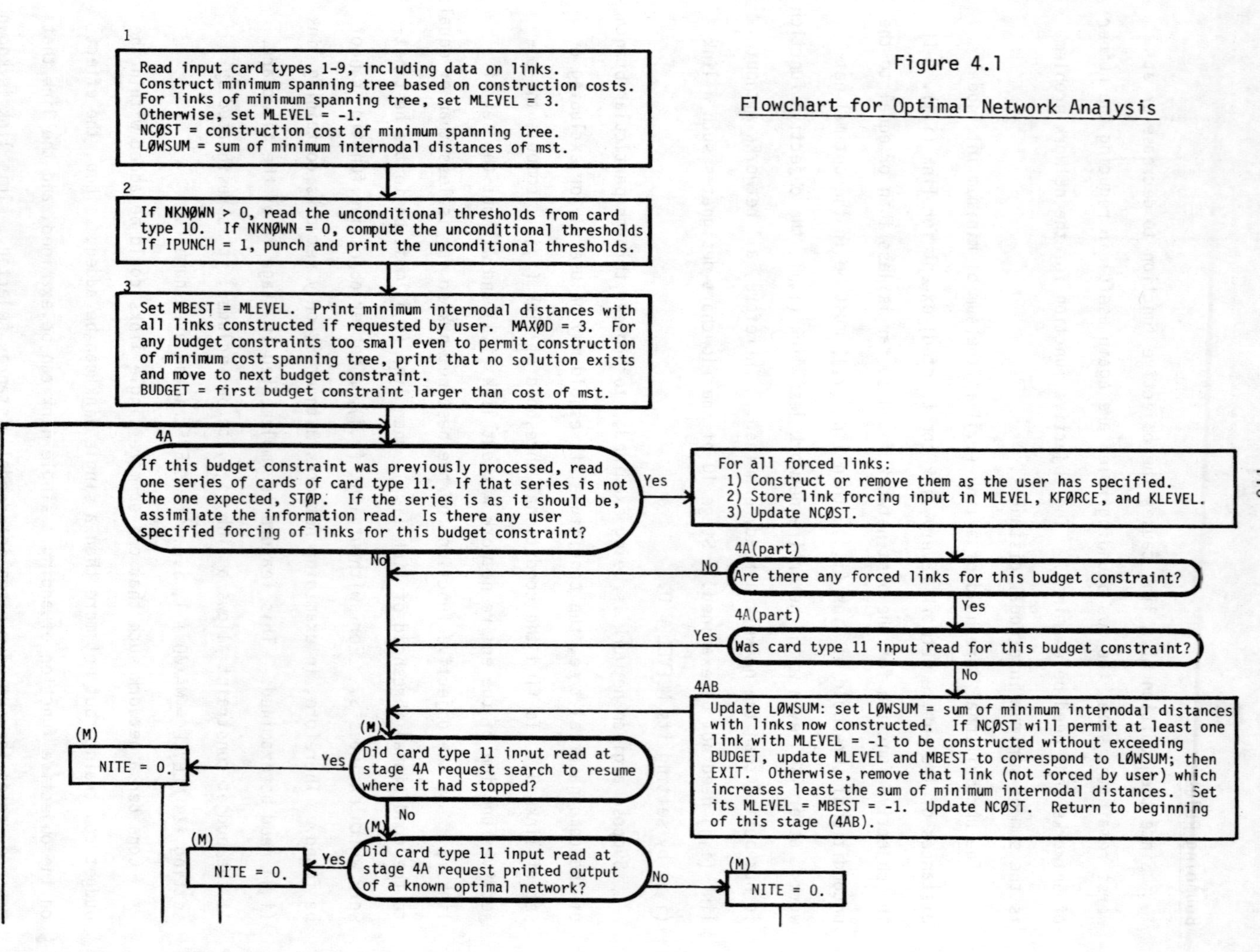

4
Of links not now constructed with MLEVEL = -1, and whose construction would not cause the network cost to exceed BUDGET, select the link whose construction most decreases the sum of minimum internodal distances.
Call the link (IFØUN1, IFØUN2).
5(M)
Was a link found at stage 4 ?
No
Yes
6
Could another link (other than one just selected) with MLEVEL = -1 be constructed within BUDGET?
Yes
No
8
Construct the link (IFØUN1, IFØUND2): add it to current network; set its MLEVEL = MAXØD + 2. NCØST = NCØST + cost of the link.
6B
Let j be any link which has 1) MLEVEL = -1, 2) is finite, and 3) can be constructed with another link at MLEVEL = -1. Let J be the link satisfying the above which among all j such links most decreases the sum of minimum internodal distances.
6C
Found a J ?
Yes
No
6D
(IFØUN1, IFØUN2) = J
7
Is the network plus the last link chosen at stage 4 the best network found yet for this budget constraint?
No
Yes
9
Record the new best network: LØWSUM = sum of minimum internodal distances of network constructed plus last link chosen at stage 4; for all links, set MBEST = MLEVEL; For the last link chosen at stage 4, set MBEST = 1. (Note: the last link chosen at stage 4 is part of best network found for this BUDGET; LØWSUM and MBEST record this. However, it is not constructed, and its MLEVEL remains -1.
10
For each link with MLEVEL > 0, set its MLEVEL = 0 if the link's unconditional threshold is ≥ LOWSUM.
11
Compute MAXØD. MAXØD = maximum odd value of the entries to MLEVEL. If that is negative, MAXOD = 1.
12(M)
MAXØD = 1 ?
Yes
No
(M)
NITE = NITE + 1.
(M)
Does NITE = user-specified iteration limit? If so, this budget constraint will be abandoned, program prints best network yet and a message saying it may not be optimal.
Yes
No
16
For each link with MLEVEL > MAXØD, remove the link from the network and set its MLEVEL = -1.
Reduce NCØST by the cost of each link removed above.
For each link with MLEVEL ≤ -MAXØD, set MLEVEL = -1.
17
Of the links with MLEVEL = MAXØD, remove link which gives least increase in sum of minimum internodal distances when removed from network. Set that link's MLEVEL = - MAXØD. NCØST = NCØST - cost of that link.
18
For each link (i,j) with MLEVEL = -1, do the following:
1) Compute its conditional threshold.
2) If its conditional threshold is ≤ LØWSUM, proceed to next link with MLEVEL = -1, and repeat (1).
3) Otherwise, if any other link (k,ℓ) which still has MLEVEL = -1 has cost ≤ [BUDGET - NCØST - cost of (i,j)], construct link (i,j), set its MLEVEL = MAXØD + 1 and set NCØST = NCØST + cost of link (i,j).
The first time a link fails the condition in (3), or if all links with MLEVEL = -1 have had their conditional threshold computed, exit stage 18.
13A
For all links user forced at this budget constraint, reset MLEVEL and KFØRCE so that these links remain forced for printing; upon entry to stage 15 these links become as any normal link. Punch specialized output if desired.
13
Construct the best network found; print whether optimal network was found, or search was halted prematurely by the user-specified iteration limit. Print any links which user forced for this budget constraint. Print all output pertaining to network found.
14(M)
Was that the last budget constraint?
Yes
STØP
No
15
For each link, if
MBEST > 0, MLEVEL = 3;
MBEST = 0, MLEVEL = 0;
MBEST < 0, MLEVEL =-1.
(M)
Set BUDGET equal to new budget constraint.

Search-ordering Procedure

Two arrays and three variables are used to record the status of the search:

1. MLEVEL is an array with n(n-1)/2 elements, one for every possible link. MLEVEL indicates what network is currently constructed and the status of the search. For a given link, if MLEVEL is:

   a) odd and greater than one, then the link is currently constructed and was selected in most cases at stage 4 or 6B; the link is not known to be in the optimal network, and therefore is a candidate for being suboptimal;

   b) even and greater than one, then the link is currently constructed and was selected by the conditional threshold criterion at stage 18;

   c) zero, assuming the link is not a special forced link which the user has designated must be constructed (or not constructed) in the desired network, then the link is constructed and must be in the optimal network by the unconditional threshold at stage 10;

   d) odd and less than -1, then the link is not currently constructed; it was of type (a) above, but has been rejected by stage 17 to examine networks without this link;

   e) -1, then the link is a candidate for construction.

2. MAXØD is a variable indicating that the search is over. MAXØD is the maximum of the odd-valued elements of MLEVEL, i.e. of type (a) above. If there are no such links, MAXØD = 1. If the search is not complete by the time that the user-specified iteration limit is reached, the program halts the search and artificially sets MAXØD = 1.

3. NCOST gives the construction cost (total length of links) of the current network as designated by the MLEVEL array.

4. MBEST is an array with n(n-1)/2 elements, one for every possible link. MBEST indicates the best network found so far in the search. For a given link, if MBEST is:

   a) positive, then the link is constructed for the best network found so far;

   b) negative, then the link is not included in the best network;

   c) zero, assuming the link is not a special forced link which the user has designated must be constructed (or not constructed) in the desired network, then the link is in the best network so far and has an unconditional threshold greater than or equal to the best objective function found so far.

5. LØWSUM is the lowest sum of minimum internodal distances found so far in the search, that is the best value of the objective function found yet.

6. Other important variables are:

   a) BUDGET, the budget constraint.

   b) NITE, the number of iterations.

The following notes help to explain the flowchart in Figure 4.1. Stage 4 uses a stepwise heuristic for finding good solutions. Of all candidate links (MLEVEL = -1), choose the one which passes the budget constraint and most decreases the sum of minimum internodal distances.

When the answer at stage 6 is NØ, the link chosen by stage 4 would not permit any more candidate links to be added to the network. Stages 6B and 6C check that there are no links with MLEVEL = -1 such that two such links could be added simultaneously within the budget constraint, if the link at stage 4 were not constructed. If there are two or more such links, select the one among those which most decreases the objective function, and return to stage 8. Otherwise go on to stage 7.

The reasoning for this addition to the general algorithm of Figure 2.1 is straightforward. Suppose upon reaching stage 7, the network being considered has K + 1 links. The algorithm is based on the result that the last link selected by stage 4 yields the optimal network given the K other links, and given that the only links considered for entry have MLEVEL = -1. The selection criterion at stage 4 guarantees this result if and only if no two links with MLEVEL = -1 could simultaneously be added to the current network. Therefore, stages 6B, 6C, and 6D guarantee that no two links with MLEVEL = -1 could be added simultaneously. With only a size constraint, as in optimal regression, this procedure is not needed.

The cutoff rule of conditional thresholds is applied in stage 18. If there are are any links with MLEVEL = -1 which could be constructed within the budget constraint upon entry to stage 18, this stage guarantees that at least one such link remains upon exit, which is the meaning of statement 3 in stage 18 of the flowchart. The purpose is as follows: if there would have been a link which stage 4 could find, stage 18 does not remove this possibility. The last link selected by stage 4 before going to stage 7 would be guaranteed to be the best, given the other links constructed.

Stage 15 reinitializes the algorithm for a new, larger budget constraint. The best network found for the previous budget constraint becomes a starting point for searching within the new budget constraint. All links not in previous best network become candidates for construction with MLEVEL = -1. All links with MBEST > 0 in previous best networks become suspect to suboptimality for new budget constraint by setting MLEVEL = 3. All links with MBEST = 0 have MLEVEL set to zero, for if their unconditional threshold guaranteed their optimality in the previous budget constraint, it will again guarantee their optimality for the larger budget constraint. (An exception to this statement about unconditional thresholds is the case of user requesting specific links to be constructed, or not constructed, for the larger budget constraint. This exception is handled in stage 4AB.)

4.3 Examples of Optimal Network Analysis

Most of the experience with the performance of the optimal network program relates to four networks of ten nodes each. The node coordinates for each network were generated by drawing ten pairs of two digit numbers from a table of random numbers; the link distances and construction costs were both taken to be the straight line distances between each pair of nodes. The node coordinates were multiplied by 10 to help insure that the distances were unique. The coordinates and internodal distances for the four networks are given in Table 4.1.

For ease of comparison of results, budget constraints were chosen as a given proportion of total network length. Initially, networks with budget constraints of 0.3, 0.4, ... 0.8 of network length were computed. A summary of the results is given in Table 4.2, which was also reported in Boyce et al. (1973). The results suggest that the four networks behave in a similar manner. For the budget constraints equal to 30 percent of total network length, the algorithm was not permitted to run to completion to conserve computer time.

Subsequently, Network 1 was chosen for more detailed examination. Networks with budget constraints of 0.1, 0.2 and 0.9 of network length were also computed. The networks for the 0.1, 0.2 and 0.3 constraints were computed in steps of 1000 iterations, and the results reviewed after each step. Computation was concluded at the limits shown in Table 4.3 showing more detailed results for nine budget constraints on Network 1. The actual links selected are shown in Table 4.4.

Table 4.5 provides some additional insight into the computational procedure, and was useful in evaluating the output of solutions for budget constraints 0.1, 0.2 and

Table 4.1

Node Coordinates, Link Distances and Construction Costs for Four Networks

Coordinates

Network	Node	X	Y
Network 1	1	420	340
	2	840	550
	3	20	10
	4	560	530
	5	780	420
	6	370	420
	7	60	210
	8	750	510
	9	550	610
	10	940	170
Network 2	1	290	590
	2	920	560
	3	500	610
	4	630	180
	5	450	250
	6	460	90
	7	320	510
	8	720	30
	9	260	860
	10	990	890
Network 3	1	130	210
	2	910	560
	3	500	310
	4	370	570
	5	860	180
	6	220	930
	7	540	690
	8	560	540
	9	280	270
	10	60	270
Network 4	1	980	890
	2	140	240
	3	760	680
	4	980	130
	5	590	690
	6	220	710
	7	480	330
	8	750	280
	9	600	910
	10	940	950

Link Distances and Construction Costs

	Network 1		Network 2		Network 3		Network 4	
Link	Length	Cost	Length	Cost	Length	Cost	Length	Cost
1, 2	470	470	631	631	855	855	1062	1062
1, 3	519	519	211	211	383	383	304	304
1, 4	236	236	533	533	433	433	760	760
1, 5	369	369	376	376	731	731	438	438
1, 6	94	94	528	528	726	726	781	781
1, 7	383	383	85	85	631	631	751	751
1, 8	371	371	706	706	542	542	652	652
1, 9	300	300	272	272	162	162	381	381
1,10	547	547	762	762	92	92	72	72
2, 3	982	982	423	423	480	480	760	760
2, 4	281	281	478	478	540	540	847	847
2, 5	143	143	563	563	383	383	636	636
2, 6	488	488	658	658	783	783	477	477
2, 7	851	851	602	602	392	392	352	352
2, 8	98	98	566	566	351	351	611	611
2, 9	296	296	725	725	694	694	813	813
2,10	393	393	337	337	898	898	1070	1070
3, 4	750	750	449	449	291	291	592	592
3, 5	864	864	363	363	383	383	170	170
3, 6	539	539	522	522	680	680	541	541
3, 7	204	204	206	206	382	382	448	448
3, 8	885	855	620	620	238	238	400	400
3, 9	801	801	347	347	224	224	280	280
3,10	934	934	564	564	442	442	324	324
4, 5	246	246	193	193	626	626	682	682
4, 6	220	220	192	192	390	390	956	956
4, 7	594	594	453	453	208	208	539	539
4, 8	191	191	175	175	192	192	275	275
4, 9	81	81	774	774	313	313	868	868
4,10	523	523	796	796	431	431	821	821
5, 6	410	410	160	160	986	986	371	371
5, 7	750	750	291	291	602	602	376	376
5, 8	95	95	348	348	469	469	440	440
5, 9	298	298	639	639	587	587	220	220
5,10	297	297	837	837	805	805	436	436
6, 7	374	374	443	443	400	400	460	460
6, 8	391	391	267	267	517	517	682	682
6, 9	262	262	796	796	663	663	429	429
6,10	622	622	960	960	679	679	759	759
7, 8	752	752	625	625	151	151	275	275
7, 9	633	633	355	355	494	494	592	592
7,10	881	881	770	770	638	638	772	772
8, 9	224	224	949	949	389	389	648	648
8,10	389	389	901	901	568	568	696	696
9,10	588	588	731	731	220	220	342	342

Table 4.2

Summary Performance Measures for Four Networks

budget constraint as proportion of total cost of network	number of links in optimal solution				number of links passing unc. threshold				number of iterations				iteration number of optimal solution			
network number:	1	2	3	4	1	2	3	4	1	2	3	4	1	2	3	4
0.3	20	20	21	20	1	1	0	0	6000*	500*	750*	300*	1	1	319	2
0.4	25	24	25	25	3	5	3	4	1329	452	2002	2439	2	1	1	13
0.5	29	27	28	28	12	14	12	11	138	224	710	968	1	1	1	1
0.6	32	31	32	32	24	21	19	17	21	90	175	120	1	1	1	1
0.7	35	34	35	35	31	28	27	28	12	17	38	14	1	1	1	1
0.8	38	38	39	39	35	33	36	37	4	9	3	5	1	2	1	2

*Algorithm stopped by iteration limit shown.

Table 4.3

Nine Solutions for Network 1

budget level	budget constraint	cost of network	objective function	number of links in optimal solution	number of links passing uncond. threshold	number of iterations	iteration number of solution
MST	-	1654	48668	9	-	-	-
0.1*	2000	1900	47708	10	0	8000	1
0.2*	4000	3929	43300	15	0	6000	3541
0.3*	6000	5888	42144	20	1	6000	1
0.4*	8000	7919	41632	25	3	1329	2
0.5	10000	9919	41408	29	12	138	1
0.6	12000	11686	41306	32	24	21	1
0.7	14000	13909	41262	35	31	12	1
0.8	16000	15683	41242	38	35	4	1
0.9	18000	17903	41234	41	40	1	1
all links	-	20619	41234	45	-	-	-

*algorithm stopped by iteration limit; solution not proved optimal.

Table 4.4

Links Constructed in Solutions for Network 1

(x denotes a link is constructed between nodes shown on lines a and b)

budget level	a	1									2								3							4						5					6				7			8		9
	b	2	3	4	5	6	7	8	9	0	3	4	5	6	7	8	9	0	4	5	6	7	8	9	0	5	6	7	8	9	0	6	7	8	9	0	7	8	9	0	8	9	0	9	0	0
MST						x										x						x					x		x	x				x		x	x									
0.1*						x										x						x				x	x		x	x				x		x	x									
0.2*			x	x	x	x	x	x		x						x						x					x		x	x				x		x								x		
0.3*			x	x	x	x	x	x		x			x			x						x			x	x	x		x	x				x		x	x		x					x		
0.4			x	x	x	x	x	x		x			x			x		x			x	x			x	x	x		x	x		x		x	x	x	x	x	x					x		
0.5			x	x	x	x	x	x	x	x			x			x	x	x			x	x			x	x	x		x	x	x	x		x	x	x	x	x	x				x	x		
0.6			x	x	x	x	x	x	x	x		x	x			x	x	x		x	x	x			x	x	x		x	x	x	x		x	x	x	x	x	x	x			x	x		
0.7			x	x	x	x	x	x	x	x		x	x			x	x	x	x	x	x	x	x		x	x	x		x	x	x	x		x	x	x	x	x	x	x			x	x		x
0.8			x	x	x	x	x	x	x	x		x	x			x	x	x	x	x	x	x	x		x	x	x		x	x	x	x		x	x	x	x	x	x	x	x	x	x	x	x	x
0.9			x	x	x	x	x	x	x	x	x	x	x	x		x	x	x	x	x	x	x	x		x	x	x		x	x	x	x	x	x	x	x	x	x	x	x	x	x	x	x	x	x

Note: node 10 is shown as 0 to facilitate column spacing.

*Algorithm stopped by iteration limit; solution not proved optimal.

Table 4.5

Order of Link Deletion from Initial Solutions for Network 1

Budget Level - Proportion of Cost of Total Network

Order of Deletion	0.1			0.2			0.3			0.4			0.5			0.6		
	link deleted	iteration number	No. of iterations	link deleted	iteration number	No. of iterations	link deleted	iteration number	No. of iterations	link deleted	iteration number	No. of iterations	link deleted	iteration number	No. of iterations	link deleted	iteration number	No. of iterations
1	3,7	2	10	1,7	2	8	4,5	2	5	7,10	2	22	4,10	2	5	1,9	2	3
2	6,7	12	24	1,5	10	79	2,5	7	10	5,6	24	37	2,9	7	7	6,10	5	5
3	5,10	36	108	1,3	89	366	3,10	17	302	3,6	61	80	6,8	14	12	4,10	10	4
4	1,6	144	195	1,8	455	1138	8,9	319	298	6,9	141	75	5,9	26	20	3,5	14	2
5	4,6	339	830	1,10	1593	1759	1,4	617	392	4,5	216	85	2,10	46	6	2,9	16	3
6	4,9	1169	1164	3,7	3352	179	1,7	1009	169	2,5	301	103	7,10	52	7	6,8	19	3
7	4,8	2333	2571	6,7	3531	1541	1,5	1178	374	3,10	404	16	5,6	59	6	last	21	
8	2,8	4904	3096+	5,10	5072	928+	1,3	1552	913	8,9	420	58	3,6	65	22			
9							1,8	2465	2191	1,4	478	181	6,9	87	13			
10							1,10	4656	548	1,7	659	70	4,5	100	13			
11							5,10	5204	250	1,5	729	36	2,5	113	8			
12							2,8	5454	546+	1,3	665	67	8,9	121	7			
13										1,8	832	361	1,4	128	1			
14										1,10	1193	22	1,3	129	1			
15										5,10	1215	2	1,8	130	4			
16										2,8	1217	11	1,7	134	5			
17										5,8	1228	8	last	138				
18										4,8	1236	19						
19										6,7	1255	44						
20										4,6	1299	31						
										last	1329							
total iterations required		> 8,000			> 6,000			> 6,000			1329			138			21	
links not yet removed:	(5,8)			(1,6) (2,8) (4,6) (4,8) (4,9) (5,8)			(1,6) (4,6) (4,8) (4,9) (5,8) (6,7)											
links passing unconditional threshold:	---			---			(3,7)			(1,6) (3,7) (4,9)			(1,5) (1,6) (1,10) (2,8) (3,7) (3,10) (4,6) (4,8) (4,9) (5,8) (5,10) (6,7)			(1.3) (1,4) (1,5) (1,6) (1,7) (1,8) (1,10) (2,5) (2,8) (2,10) (3,6) (3,7) (3,10) (4,5) (4,6) (4,8) (4,9) (5,6) (5,8) (5,9) (5,10) (6,7) (6,9) (7,10) (8,9)		

0.3. To find an initial solution, first the program finds the minimum spanning tree (MST) and then adds links in a stepwise manner such that the link resulting in the largest decrease in the objective function is added at each step. Of course, only links that fit within the available budget are considered. These links are constructed, or included in the network, at MLEVEL = 3. Subsequently, each of these links is deleted from the network, and all possible networks excluding this link are considered. Hence, the deletion point of one of these initial links is a major milestone in the computation. The first link chosen for deletion is the one which results in the least increase in the objective function, etc. When the last link with MLEVEL = 3 is deleted, the algorithm is nearly finished, having only to consider all possible networks excluding that link. Table 4.5 shows the order in which links were deleted from Network 1, and the iteration number on which deletion occurred. In the case of solutions for budgets 0.1, 0.2 and 0.3, the links yet to be removed are also shown. Links passing the unconditional threshold are shown at the bottom of the table. These links need not be deleted, greatly reducing computation time.

More detailed results for Networks 2, 3 and 4 are shown in Tables 4.6, 4.7 and 4.8. In all of the results reported here, the network for each budget constraint was computed with the minimum spanning tree as the initial solution. The program does permit the use of the solution for a previous, smaller budget constraint as the initial solution. However, this procedure does not result in an objective comparison in the number of iterations required. In addition, limited experience suggests fewer iterations are required if the MST initial solution is used.

4.4 Suggestions for a Strategy for Using the Program

The optimal network program finds the network that minimizes the sum of the minimum path distances between all pairs of nodes, and satisfies the budget constraint. Obviously, the total number of networks to be considered is quite large, and enumerating each one to find the best would take very long. The program uses cutoff rules based on thresholds to reduce greatly the number of networks that need be examined.

The cutoff rules essentially rate the links as to their usefulness. Consequently, a network with certain links essential for short minimum paths can be completed quite

Table 4.6

Solutions for Network 2

budget level	budget constraint	cost of network	objective function	number of links in optimal solution	number of links passing uncond. threshold	number of iterations	iteration number of solution
MST	-	2141	65852	9	-	-	-
0.3	6900	6873	47320	20	1	500*	1
0.4	9200	9113	46832	24	5	452	1
0.5	11500	11413	46886	27	14	224	1
0.6	13800	13699	46462	31	21	90	1
0.7	16100	16023	46396	34	28	17	1
0.8	18400	18340	46374	38	33	9	2
all links	-	23182	46364	45	-	-	-

*algorithm stopped by iteration limit.

Table 4.7

Solutions for Network 3

budget level	budget constraint	cost of network	objective function	number of links in optimal solution	number of links passing uncond. threshold	number of iterations	iteration number of solution
MST	-	2183	61680	9	-	-	-
0.3	6600	6592	45368	21	0	750*	319
0.4	8800	8757	44710	25	3	2002	1
0.5	11000	10958	44392	28	12	710	1
0.6	13200	13130	44238	32	19	175	1
0.7	15400	15390	44150	35	27	38	1
0.8	17600	17540	44104	39	36	3	1
all links	-	22044	44088	45	-	-	-

*algorithm stopped by iteration limit.

Table 4.8

Solutions for Network 4

budget level	budget constraint	cost of network	objective function	number of links: in optimal solution	number of links: passing uncond. threshold	number of iterations	iteration number of solution
MST	-	2415	66936	9	-	-	-
0.3	7500	7367	52034	20	0	300*	2
0.4	10000	9940	51052	25	4	2439	13
0.5	12500	12199	50738	28	11	968	1
0.6	15000	14929	50508	32	17	120	1
0.7	17500	17189	50394	35	28	14	1
0.8	20000	19892	50340	39	37	5	2
all links	-	25161	50322	45	-	-	-

*algorithm stopped by iteration limit.

quickly with only a few networks examined. However, if many links have a similar importance for short minimum paths, the cutoff rules are inefficient. Therefore, sometimes the search will proceed very slowly because of the great number of networks examined. The differences in the sum of minimum internodal distances may be small for networks with many links, or combinations of links, which add little to the network. Of course, the larger the number of nodes, the greater the problem becomes.

Consequently, the program provides a procedure to control the amount of computation. At stage 12 of the flowchart, a test is made to determine whether the program has finished its search for the optimal network. Each time the test is performed is counted as an iteration of the algorithm. The user may specify a limit to the number of iterations for any given budget constraint. Essentially, then, the user can limit the number of networks examined for a given budget constraint.

This procedure, of course, may result in suboptimal networks. However, as shown in Section 4.3, the best solution tends to be the initial solution or an immediate improvement on the initial solution, at least for the simple networks considered to date. In addition, if suboptimal networks for several budget constraints are computed sequentially, the degree of suboptimality could be compounded as the sequence progresses. Not enough computational experience is presently available to assess the extent of this problem.

Because user specified iteration limits may cause the program to find suboptimal networks, the user can request punched output for restarting the search at a later date. The output is punched when the search ends normally, or when the search is halted by the iteration limit. If the search ends normally and punches an optimal result, it can be read at a later time to print results for several networks together; little recomputation is necessary. If the search was halted by the iteration limit, reading in the punched output at a later date causes the search to be resumed at the exact point where it was halted.

For details on specifying iteration limits, examine card type 4. For specifying punched output and the form of punched output, see IPUNCH on card type 2 and all of card types 10 and 11. For reading in punched output, see card types 5, 10 and 11.

The user may also specify that certain links be considered in special ways:

1. force a link to be included (constructed) in every network for every budget constraint;
2. force a link to be included for a particular budget constraint;
3. force a link to be excluded for a particular budget constraint.

This capability can be very useful. For instance, suppose a network exists in reality, and certain links are proposed for new construction. It is desired to identify the optimal extension to the existing network. The user can force the existing links to be constructed with options 1 or 2 above. If those links should be constructed for every budget constraint, use option 1. Then, the construction cost could be set to zero on the link cards (card type 9) and the budget constraint can be the actual amount available for new construction. However, if one of the networks desired has a different set of links to be forced, assign the real construction costs for all links on card type 9, set budget constraints taking into account these added construction costs, and use option 2 above. For details on forcing links, examine IFØR of card type 2 and card type 6.

4.5 Order and Detailed Description of Input Card Types

The input cards are arranged in the following order:

1. Title
2. Problem definition
3. Budget constraints
4. Iteration limits
5. Networks previously processed (optional)
6. Links to be forced (optional)
7. Output options
8. Format of links
9. Deck of links
10. Unconditional thresholds (optional)
11. Networks from previous jobs (optional)

The program can process several problems in one job: set up the first problem as shown above, followed by card types 1-11, as needed, for the second problem, etc. After finishing a problem, the program will try to read a new title card, and hence execution continues until there is no input.

4.5.1 Title - FØRMAT(20A4)

Any 80 characters may be used to identify the problem.

4.5.2 Problem Definition - FØRMAT(7I5)

NØDES - number of nodes in the network

NMAX - total number of budget constraints

NLINKS - number of links to be read from cards

NKNØWN - indicates whether or not parts of the problem have been worked on before; results are to be read on card types 10 and 11. See Section 4.4 for an introduction to using this option; it is used chiefly if an iteration limit halted the search before completion during a previous job.

0, normal; no punched output from an earlier job to be read.

i, where i > 0 means i of the NMAX budget constraints were worked on before; card type 10 and i sets of card type 11 will be read.

Note that setting NKNØWN > 0 has a minor effect on the output options card, card type 7. One cannot obtain as initial output, before starting the search for optimal solutions, the matrix of minimum distances with every link constructed. This is noted under option IALL on card type 7 which controls the printing of that matrix. This inconvenience is not unreasonable since NKNØWN > 0 is essentially a restarting procedure, and therefore the matrix could have been printed in the earlier job.

IPUNCH - 0, normal; punch no output.

1, punch special output to be read as card types 10 and 11 in a subsequent job. If NKNØWN > 0, in a previous job, IPUNCH must have been set to 1.

IFØR - 0, normal; no finite length links are to be forced to be included nor excluded from the possibility of construction.

j, where j > 0 indicates j links are to be forced. Each of the j links may be forced to be included for some budgets, or excluded for some budgets, or both, or forced to be included for all budget constraints. See Section 4.4 for an introduction.

IØ - 0, normal, print the normal user output.

1, print intermediate output. THIS IS FOR THE PURPOSE OF DEBUGGING THE PROGRAM ONLY. It prints the values of important arrays and variables subroutine by subroutine and traces flow through the program. Even for networks as small as 7 nodes, the output is extremely large.

4.5.3 Budget Constraints - FØRMAT(8I10)

For every budget constraint, the program will find the optimal network which has total construction cost less than or equal to the budget constraint, unless the user specified iteration limit halts the search prematurely. The budget constraints must be in ascending order. There must be NMAX budget constraints. The values are read into array NSIZES. One at a time, they are selected and stored in BUDGET. If there are more than 8 budget constraints, place constraints 9 through 16 on a second card, and so on.

4.5.4 Iteration Limits - FØRMAT(8I10)

An iteration limit sets a limit on the number of networks examined for a budget constraint. An iteration limit is needed for each budget constraint, or NMAX in all; they are read into array NITL. The first field on the card corresponds to the first budget constraint. If NMAX > 8, continue limits for 9-16 on a second card, and so on.

4.5.5 Networks Previously Processed - FØRMAT(16I5)

This card is used if and only if NKNØWN > 0 on card type 2. See Section 4.4 for an introduction to the use of this card.

This card type is used only when partially complete or complete optimal networks are available on punched cards from previous jobs. Unless a previous search was stopped before the optimal network was found, it is unlikely that this capability would be used.

The budget constraints are numbered sequentially beginning with the first constraint on card type 3. Card type 5, giving the networks previously processed, lists the network number as defined above. The purpose is to identify which budget constraints have punched output to be read in as card type 11. There must be NKNØWN of these values, where NKNØWN is defined on card type 2; the values must be in ascending order. Values are read into array IKNØWN. If NKNØWN > 16, continue values 16 through 32 on a second card.

For an example, consider a problem with 15 nodes. In a preliminary, exploratory job, networks were requested with budget constraints of 4000 and 10000. The results with 4000 stopped because of its iteration limit, but the network for 10000 was proven optimal by the program. Now the user desires networks of 3500, 5000, 7000, and 9000 to see the differences in sets of links chosen. He also wants to have the optimal result for 4000. To have all the results in one set of output, he will also read in the result for 10000 which will be printed without a search. The user had requested punched output during the job with 4000 and 10000 by setting IPUNCH = 1 on card type 2. To set up the new problem using the previous results, he sets NMAX = 6, NKNØWN = 2 on card type 2. NKNØWN = 2 because information on two networks with budgets 4000 and 10000 is to be read. Since punched output is not needed this time, IPUNCH = 0. The budget constraints card has 3500, 4000, 5000, 7000, 9000, 10000. Card type 5 has values of 2 and 6 in columns 5 and 10 respectively; the setup is illustrated below:

Network Number	Budget Constraint	
1	3500	not processed before
2	4000	partially complete solution to be read
3	5000	not processed before
4	7000	not processed before
5	9000	not processed before
6	10000	completed solution to be read

The program will expect card type 11 for budgets 4000 and 10000. Also, card type 10 is expected since NKNØWN > 0.

4.5.6 Links to be Forced - FØRMAT(12I5)

Include only if IFØR > 0 on problem definition card. The purpose of card type 6 is to designate which links are to be forced for which budget constraints. Each forced link corresponds to one and only one card of this card type. Therefore, there must be IFØR of these cards. The kth card is read into row k, columns 1-12 of array KLEVEL.

To designate that a link is to be forced place its two node numbers in the first two fields of five columns on the card in either order. If the link is to be constructed for every budget constraint, place a 100 in the third field of 5 columns. Otherwise, every budget constraint must be identified for which the link is to be forced using fields 3 through 12 as follows:

1. If a link must be included for the jth budget constraint, place the integer j in the next unused field of 5 columns;

2. If a link must be excluded for the jth budget constraint, place (-j) in the next unused field of 5 columns.

The following example illustrates this procedure; five links are to be forced, and therefore five cards are used.

	link		budgets		
column:	5	10	15	20	25
card 1:	4	1	100		
card 2:	1	3	-2	1	
card 3:	2	4	1	2	3
card 4:	3	5	1	-2	
card 5:	2	6	100		

Suppose that there are 4 budget constraints: 1200, 2000, 2800, and 3500. The effect is as follows:

1. For budget of 1200, links (1,3) (2,4) and (3,5) will be included (constructed), as each link has (+1) in columns 15 or 20. Links (1,4) and (2,6) will be included because of the 100 in the third field.

2. For budget of 2000, links (1,3) and (3,5) will be excluded from consideration because of the (-2) on cards 2 and 4. Link (2,4) will be constructed because of the (+2) on its card. Links (1,4) and (2,6) will also be constructed.

3. For budget of 2800, link (2,4) will be constructed because of (+3) on its card, and links (1,4) and (2,6) will be constructed as above.

4. For budget of 3500, links (1,4) and (2,6) will again be constructed.

Note that the cards may be in any order, the pair of nodes may be in either order, and the identifiers of budget constraints may be given in any order.

The optimal network program always works with a set of links such that its total cost permits only one additional link to be constructed. It then chooses the best additional link given the previous set. Therefore, in forcing links for some budget constraint, the total cost of these links must be small enough to permit at least some other link to be constructed without the budget constraint being exceeded. Otherwise, the program will stop, print the cost of present network, the budget constraints, and messages declaring the situation hopeless because the user has forced too many links.

If the user forces links and requests punched output (IPUNCH = 1), then for any future job where the punched output is read as card type 11, the forced links must again be indicated on card type 6.

4.5.7 Output Options - FØRMAT(5I5)

LINKPR - 0, do not print the link data (card type 9)

1, print the link data

ISPAN - 0, do not print minimum internodal distances of minimum spanning tree

1, print minimum internodal distances of minimum spanning tree

IALL - 0, do not print the minimum distances with all links

1, print the minimum internodal distances with all links constructed. Note: If using the restarting procedure (NKNØWN > 0 on card type 2), then do not set IALL = 1, for the matrix printed will be incorrect. Only use IALL = 1 when NKNØWN = 0; the matrix printed will always be correct then.

NØDBAK - 0, do not print back node matrix

1, for each budget constraint, print the first back node matrix corresponding to the minimum internodal distances of the chosen network.

MØUT - 0, normal option; do not print MLEVEL array just before entering stage 11 of flowchart.

1, print MLEVEL array whenever entering stage 11 from stage 7 or stage 10 of flowchart. Once one is quite familiar with the flowchart of the algorithm, the printing of array MLEVEL can yield much information about how the search is proceeding; see for example Table 4.5. To identify a given link in the MLEVEL array, determine the link's sequence number in the ordered list of links: (1,2),(1,3) ... (1,NØDES), (2,3),(2,4) ... (2,NØDES)... (NØDES-1,NØDES). The corresponding element in MLEVEL is the status of the desired link.

4.5.8 Format of Links - FØRMAT(20A4)

The link data can have almost any format, because the user specifies the format. The format statement should provide for the reading of one and only one link. All fields must be integer (I) format. To prepare this card, follow the rules for FØRTRAN IV variable format cards. Remember that the first character of the format must be a left parenthesis, and the last character must be a right parenthesis. The card is read into array FMT.

4.5.9 Deck of links - User Specified Format

Since the format for reading the links provides for reading one and only one link, one card must be used for each link. There must be NLINKS links, where NLINKS was given on the problem definition card. If NLINKS < NØDES(NØDES-1)/2, all links not in the deck will be assumed not to exist and given infinite length and infinite construction cost. Such links will never be considered by the program.

The data for a link should be given in the following order:

1. field 1 - first node
2. field 2 - second node
3. field 3 - length
4. field 4 - construction cost

All links are defined to be two-way links. The nodes may be in either order, and the links cards may be arranged in any order.

The program can handle networks of fifty nodes maximum. Link lengths and construction costs may assume any integer value i, where $0 \leq i \leq 17{,}119$. Link lengths are stored above the diagonal of the NØDES by NØDES matrix LD. Construction costs are stored below the diagonal of LD.

4.5.10 Unconditional Thresholds

Use this card only when NKNØWN > 0 on card type 2. This card type was punched during a previous job by setting IPUNCH to one. This card type is part of the punched output to facilitate restarting computation halted by iteration limits; see Section 4.4 and NKNØWN and IPUNCH on problem definition card, Section 4.5.2.

The cards making up this card type are the following:

1. The first card is a label for the convenience of the user. "UNCØNDITIØNAL THRESHØLDS -- NETWØRK" are punched in columns 2-36.
2. Unconditional thresholds for all links are punched with FØRMAT(8I10) in ascending order: (1,2),(1,3) ... (1,NØDES) ... (NØDES-1,NØDES).

Unconditional thresholds are an important characteristic of the links, which enable the program to reduce the total number of networks examined in the search for the optimal network. The numerical value of the unconditional thresholds is unique to the set of finite links and their link lengths. If a formerly nonexistent link (infinite length and cost) is added, or if a finite link is deleted (given infinite length and cost), or if any link length is changed, the unconditional thresholds must be recomputed; reading in old ones could give incorrect results. This note of caution does not apply to the capability of forcing links, as the program handles them in a different way.

4.5.11 <u>Networks from Previous Jobs</u>

Use this card type only when NKNØWN > 0. This card type was punched by the program in a previous job when the user set IPUNCH = 1. Card type 11 consists of a series of cards. When IPUNCH = 1, the program punches the series for every budget constraint specified in that problem. The cards within a series are punched in correct order by the program. Once the user has selected the budget constraints for which he wants the search for optimal network to be completed, or for which he wants the optimal network reprinted, he need only find each series of cards and include them as card type 11. For every value on card type 5, there must be a corresponding series of cards of type 11. The order on card type 5 and the position of every series of card type 11 must be identical, that is, the budget constraints must be increasing. There must be NKNØWN of these series of cards. Together, the NKNØWN series make up card type 11.

A series of cards consists of the following:

1. Identifier - FØRMAT(I5,I10,I5)

   N2 - value of NØDES

   N3 - value of the current budget constraint

   N4 - 0, search for optimal result should be continued because iteration limit stopped search.

   1, optimal network should be printed without any search for it, because this network was proven optimal. After N4, the card has punched on it "NØDES, BUDGET, IØPT MBEST" if N4 = 1, or "NØDES, BUDGET, IØPT MLEVEL THEN LØWSUM MBEST" if N4 = 0.

   If N2 ≠ NODES, or if N3 ≠ current constraint, the program will print the values of the three variables, write "ERRØR -- DECK IS MIXED UP", then write that at least one of the pairs is not a match, and stop all execution. To correct the problem, locate the error in the order of the series of card type 11, comparing them with the order on card type 5.

2. MLEVEL - FØRMAT(20I4)

   MLEVEL in general identifies what links are currently constructed and identifies the location of the search for the optimal network when the search stopped. The MLEVEL values are not punched if N4 = 1.

3. LØWSUM - FØRMAT(1X,I10)

   This card is punched only when N4 = 0. The variable is the sum of the minimum internodal distances for the best network found yet.

4. MBEST - FØRMAT(20I4)

   This array is always punched. The MBEST array is a set of NØDES(NØDES-1)/2 integers, one for each node pair (i,j), where $i < j$, and in the same order as MLEVEL. MBEST indicates which links are constructed in the best network thus far.

The purpose of the series of cards is to permit a search for the optimal network to resume at the very point where it was stopped by iteration limits or to permit previously computed optimal results to be printed in the same output with new results, but without searching again for the optimal result.

4.5.12 WARNING!

There are essentially no checks on the input. Therefore, one could specify meaningless directions and get unpredictable, incorrect results. The user should check the input very carefully. For instance, one could set NØDES $\leq$ 0 or NLINKS $\leq$ 0 or NKNØWN $<$ 0. It is conceivable that an infinite loop could result. Admissible values of input are as follows:

1. Title - maximum of 80 characters; if no title is wanted, include a blank card.

2. Problem definition

   a) 0 $<$ NØDES $\leq$ 50

   b) 0 $<$ NMAX $\leq$ 50

   c) NLINKS $\leq$ NØDES(NØDES-1)/2 There must be at least enough links to construct a connected network.

   d) 0 $\leq$ NKNØWN $\leq$ NMAX

   e) IPUNCH = 0 or 1

   f) 0 $\leq$ IFØR $\leq$ 30

3. Budget constraints can be less than the cost of minimum cost spanning tree. The program would print that there is no solution. However, the budget constraint should never exceed the total cost of a complete network by more than the maximum allowed link cost, 17,119.

4. Iteration limits must be a nonnegative integer less than 2,147,000,000.

5. Networks previously processed: every value i must have $0 < i \leq$ NMAX.

6. Links to be forced: every node number n must have $0 < n \leq$ NØDES. The third field, k can be 100 or any value $0 < |k| \leq$ NMAX; for all additional fields, $0 < |j| \leq$ NMAX.

7. Output options

   a) LINKPR = 0 or 1.

   b) ISPAN = 0 or 1.

   c) IALL = 0 or 1.

   d) NØDBAK = 0 or 1.

   e) MØUT = 0 or 1.

8. Format of links: no more than 80 characters may be used to specify format of link data.

9. Deck of links: each node number n must have $0 < n \leq$ NØDES. For link length ℓ, $0 \leq \ell \leq 17119$; for construction cost k, $\overline{0} \leq k \leq 17119$.

10. Unconditional thresholds - punched by program.

11. Networks previously processed - punched by program.

4.6 Output of the Program

First, the program prints the title and values of card types 2, 3, and 4. The program numbers the networks sequentially in the output, and prints the network numbers along with the budget constraints and iteration limits in a table. If card type 5 is read, its values are printed under the heading "NETWØRK NUMBERS ØF THØSE WHICH WERE PREVIØUSLY PRØCESSED." The contents of each card of type 6 are printed as follows: "LINK (field 1 , field 2) IS FØRCED FØR THE FØLLØWING NETWØRKS" followed by fields 3 through 12. Note that any blank field in card type 6 will be printed as zero.

Next, the contents of card type 7 are printed with the format for reading the links. If LINKPR = 1 on card type 7, the links will be printed when read with their length and construction cost. If a link is mispunched with identical origin and destination nodes, the error is printed and the program stops execution.

The minimum spanning tree on the basis of construction cost is printed next as a matrix. Entry (i,j) corresponds to a link from i to j. An entry of ***** means the link is not constructed. Numerical entries gives the construction cost of the constructed links. Entries (i,i) are printed as zero. The minimum internodal distances for the minimum spanning tree are printed if ISPAN = 1 on card type 7. Regardless of the value of ISPAN, the two following headings and values are printed for the minimum spanning tree:

1. "SUM OF MINIMUM INTERNØDAL DISTANCES ØF MINIMUM SPANNING TREE"

$$\sum_{i=1}^{NØDES} \sum_{\substack{j=1 \\ j \neq i}}^{NØDES} md(i,j)$$

where md(i,j) is the minimum distance between nodes i and j.

2. "CØST OF MINIMUM SPANNING TREE"

$$\sum_{i=1}^{NØDES-1} \sum_{j=i+1}^{NØDES} cost(i,j)$$

where cost(i,j) = 0, if link (i,j) is not constructed;
cost of link (i,j) if constructed in the minimum spanning tree.

If IALL = 1 on card type 7, and NKNØWN = 0 on card type 2, the minimum internodal distances for the network with all links constructed is printed. However, recall that IALL = 1 and NKNØWN > 0 causes an incorrect matrix to be printed; see IALL description in Section 4.5.7. If IPUNCH = 1 on card type 2, the unconditional thresholds are printed and punched.

If the first few budget constraints are so small as to not permit the minimum spanning tree to be constructed, the program writes for each such budget constraint: (a) title; (b) message "THERE IS NØ SØLUTION;" (c) budget constraint; and (d) construction cost of the minimum spanning tree.

While searching for the optimal network, four types of output can occur: (a) printing of networks under consideration; (b) printing of array MLEVEL, if MØUT = 1 on card type 7; (c) printing of information being punched on cards; and (d) printing of the final selected network. In the search for the optimal network, many networks are checked. Every network which is a candidate for the optimal network will be printed. If the candidate is not the best thus far, a message "WILL NØT BE SELECTED" is printed along with its sum over all node pairs of the minimum internodal distances and the best sum of minimum internodal distances thus far. If the candidate network is the best of any network found for this budget constraint, the network is printed with a message, "THE BEST SET ØF LINKS FØUND THUS FAR", its sum of minimum internodal distances and the previous best sum of minimum internodal distances. The array MLEVEL is printed immediately after the above output if MØUT = 1 on card type 7; see definition of MØUT, Section 4.5.7 for a description of this output.

If IPUNCH = 1 on card type 2, the program prints output as it punches for each budget constraint. The output states this is STAGE 13A and includes: (a) MLEVEL array, if it is punched; (b) value of LØWSUM, lowest sum of minimum internodal distances, if it is punched; and (c) MBEST array. The format for the printing is roughly the same as the punched cards.

The final output for each budget constraint consists of the title and descriptive information for the chosen network. First, the title and budget constraint are printed. If the network was proven optimal, the message "THIS IS AN ØPTIMAL NETWØRK" is printed. If the search was halted by an iteration limit, the message "SEARCH STØPPED BY REACHING ITERATION LIMIT. ØPTIMALITY NØT GUARANTEED" is printed. If any links are forced by user specification to be included or excluded for this budget constraint, these links are printed along with a warning that better networks may exist without such forcing.

The sum of minimum internodal distances and the total construction cost as defined above are printed. Also, the sums of the rows of the minimum distance matrix are printed as an measure of network connectivity. To show which links are constructed, a matrix is printed whose (i,j) entry is ***** if link (i,j) is not constructed, and the link length if link (i,j) is constructed for this budget constraint. Next the matrix of minimum internodal distances is printed. If NØDBACK = 1 on card type 2, the first back node matrix is printed corresponding to the minimum distances.

4.7 Machine Dependent Features and Suggestions for Modification

The program was written in FØRTRAN IV on an IBM 360/75. The only incompatibility with other FØRTRAN IV compilers concerns the length of INTEGER and DØUBLE PRECISIØN variables. A DØUBLE PRECISIØN variable is assumed to use exactly twice the number of words of storage as an INTEGER variable. The assumption is used in the following situation. Throughout most of the program the labeled CØMMØN block MATRIX contains F(50,50), B(50,50), ID(50,50), MD(50,50), R(50,50), which are all INTEGER arrays. However, in subroutine SPAN, the COMMON block MATRIX contains DISTNZ(50,50), ID(50,50), MD(50,50), R(50,50), where DISTNZ is DØUBLE PRECISION and ID, MD, and R are INTEGER. This of course assumes that DISTNZ(50,50) requires exactly the same storage as F(50,50) and B(50,50) combined. This may not be the case for other computers; one should note that at this point in the program, the contents of F, B and ID are unimportant. One could use filler, dummy variables to force MD and R to be in proper position in subroutine SPAN. Note that MD also could be ignored if necessary in SPAN; however, MD would have to be reinitialized again to MD(i,j) = INFIN, for $i \neq j$; and MD(i,i) = 0 for all i after CALL SPAN in subroutine SETUP.

The program requires approximately 28,000 words of core for array storage and 20,000 words for object code with the limits given in Section 4.5.12.

To change the maximum number of nodes from 50 is an involved procedure. First, determine the maximum positive integer which may be stored in a FØRTRAN INTEGER word, and call it ℓ. Let int(X) represent the function which finds the greatest integer less than or equal to X. If the program is to have a maximum number of nodes n, then link lengths and construction costs must have nonnegative integer values less than or equal to $\text{int}\left(\ell/n(n-1)^2\right) - 1$, in order that n does not become so large that the range of admissible costs or lengths is too restrictive. Given than n is chosen such that the admissible cost and length values are satisfactory, change the statement in the MAIN program assigning the integer constant to INFIN from INFIN = 838859 to $\text{INFIN} = \text{int}\left(\ell/n(n-1)\right)$.

Also, increase the dimension of all 50 by 50 square matrices to n by n. These matrices are KFØRCE, F, B, ID, MD, R and LD; all are in three labeled CØMMØN blocks which appear in many subroutines. Similarly expand to n by n the matrices D and P which are local to subroutine LINKIN and the matrices DP and PP which are local to subroutine LINKOUT. Matrix DISTNZ in subroutine SPAN must then be increased to n by n, but since it shares the same space as matrices F and B, this increase will not add to the storage requirments. Also, in subroutine SPAN, matrix NCØN must be increased to n by 2.

Several vectors need to be increased from length 1225 to length n(n-1)/2: (a) MLEVEL, MBEST, and LTHRES occur in several subroutines and in two labeled CØMMØN blocks: (b) LVP1 and LVP2 occur in labeled CØMMØN in subroutines CHKVAR and STAJ18; (c) arrays CT and MSAVE occur in subroutine STAJ18. Also, increase the vectors M and N from length 50 to length n. They occur in labeled CØMMØN in several subroutines. All such vectors and matrices which would have to be expanded are INTEGER except for matrix DISTNZ whose expansion does not have to be calculated since DISTNZ shares space with F and B.

If the range for the link costs and link lengths need to be changed, the procedure is also involved. Recall that link lengths and costs must be integers. The only way to increase the allowable range is to decrease the maximum number of nodes the program can handle; see the above paragraphs. The reasoning is simple. Because the computer must use a finite integer to represent infinite link length and cost, the range of link lengths and costs must not permit a finite length to reach the integer representing infinity (INFIN). Let n be the maximum number of nodes the program can handle and ℓ the maximum link length or construction cost. $(n-1)\ell$ must be < INFIN so that no finite path appear infinite. Let k be the maximum positive value of a FØRTRAN integer. Then, if $n(n-1)INFIN \leq k$, overflow will not occur. From these calculations follow the definition of INFIN and the permissible range on link lengths and costs.

Changing the limit of 50 on the number of budget constraints, NMAX, requires relatively little effort. Assume the new limit is to be n. Increase the vectors IKNØWN, NSIZES and NITL from 50 to n. These occur separately in three labeled CØMMØN blocks in several subroutines.

To change the limit of 30 links specified as forced, let the new limit be k. One of the first executable statements in the MAIN program is LIMFØR = 30. Change this to LIMFØR = k. Change the dimension of matrix KLEVEL from 30 by 16 to k by 16.

KLEVEL appears in the labeled CØMMØN block SPCLNK and in several subroutines as well as the MAIN program.

4.8 Description of the Program by Subroutine

This section is chiefly for those who need to understand the coding of the program or the debugging output obtained when IØ is set to 1 on card type 2. If additional or new options are to be made, this section should be very helpful. It could also be helpful in expanding or decreasing the limits on program parameters such as the number of nodes. The flowchart should be used concurrently with this description.

The program uses many square arrays of dimension NØDES by NØDES, such as F, B, ID, MD, R, KFØRCE, D, P, DP, PP. In these arrays, the link or path from node i to node j is associated with entry (i,j). For instance, the current minimum internodal distances are stored in MD, so MD(i,j) gives the current minimum path distance between node i and node j. Also, R contains the current back node matrix for minimum paths; thus, R(i,j) gives the back node of minimum path from node i to node j.

LD and DISTNZ are also NØDES by NØDES arrays. However, the upper triangular matrix stores a quantity unrelated to the lower triangular matrix. For $i < j$, LD(i,j) gives the link length, and LD(j,i) gives the link's construction cost. LD is an INTEGER array; DISTNZ is an associated DØUBLE PRECISIØN array with:

$$DISTNZ(i,j) = LD(i,j) + (i - 1)\varepsilon + j\varepsilon$$

where $\varepsilon = 0.0003$. DISTNZ therefore is the same as LD except for a slight perturbation to insure that all link lengths and construction costs are unique; however, the following equality holds:

$$LD(i,j) = \mathrm{int}\big(DISTNZ(i,j)\big)$$

where int is the "greatest integer contained in" function.

Several one-dimensional arrays are also used to store upper triangular matrices such as link lengths. In such arrays, if n = NØDES, then the sequence order of links is (1,2), (1,3) ... (1,n), (2,3), (2,4) ... (2,n) ... (n-2,n-1), (n-2,n), (n-1,n). Note that the (i,i) entry of each row is not included. To retrieve link (j,k), $j < k$, from such an array, find the following element:

$$\frac{n(n-1)}{2} - \frac{(n-j)(n-1-j)}{2} - n + k$$

The array has n(n-1)/2 elements, one for each link. Important vectors using this storage mechanism are MLEVEL, MBEST, MSAVE, LTHRES, and CT, all INTEGER arrays. When searching among all links for a subset of links with certain characteristics, the program uses two DØ loops of the form:

```
    N1  =  NØDES - 1
    INDEX  =  0
    DØ ×× I = 1,N1
    IP = I + 1
    DØ ×× J = IP,NØDES
    INDEX = INDEX + 1
 ×× Statement concerning MLEVEL(INDEX)
      (or one of the other arrays defined above)
```

Whenever the sum of minimum internodal distances is computed, it is a sum over all (i,j), i = 1 ... NØDES, j = 1 ... NØDES. Since this is a symmetric network with main diagonal elements equal to zero, it is computed as twice the sum over all (i,j), i < j. In computing the construction cost of a partial or complete network, however, the sum of link construction costs is taken over all (i,j), i < j, but not multiplied by two. These two definitions are important.

4.8.1 MAIN

The most significant purpose of MAIN is to direct flow through the algorithm. Almost all of the stages on the flowchart are separate subroutines, and most of the subroutines are called by MAIN; stages performed in MAIN are indicated on the flowchart by "(M)" after the stage number. In addition, MAIN reads card types 1-6, performs part of stage 1 and the latter part of stage 3.

Several variables are defined here and do not change throughout the program:

INFIN - integer used as the infinite link length or infinite link construction cost, presently defined as 838859. If K is the maximum positive fixed point number the computer can store, and L is the maximum number of nodes the program can accept, then INFIN = $\text{int}(K/L(L-1))$. Maximum link length and construction cost is then $\text{int}(K/L(L-1)^2) - 1$.

LIMFØR - maximum number of links the program can force, which is the number of rows in matrix KLEVEL; at present, LIMFØR = 30.

NØDES - number of nodes

NLINKS - number of links read from cards

N1 - NØDES - 1

NL - number of possible links, NØDES(NØDES-1)/2

After reading card type 1, arrays KFØRCE and KLEVEL are set to zero. Immediately after the call to subroutine SETUP, array F contains the minimum internodal distances with all links constructed. After the return from subroutine SETUP, the diagonal of arrays F, B, ID, and MD (all INTEGER arrays) are initialized. Since F, ID, and MD contain minimum distances, their diagonals are set to zero. However, for the first back node matrix, B, B(I,I) = I.

After executing subroutine STAGE2, MAIN executes the part of stage 3 that sets MBEST = MLEVEL. The statements "125 MAXØD = 3" through "GO TO 128" are the end of stage 3. These set MAXØD, store the first budget constraint in BUDGET, an INTEGER variable, and then search for the first budget constraint that at least permits the minimum spanning tree to be constructed. If some budget constraints do not permit construction of the minimum spanning tree, the following debugging output can be printed: the next budget constraint (BUDGET) after the constraint just considered, an index indicating the number of the next budget constraint (MLIST), and the array of budget constraints (NSIZES).

After returning from stage 4A, N4 = 0 means that stage 4A has read in a network from card type 11, and the search for the optimal network was not complete. Consequently, the search should resume with the network read in. N4 is set to one, so that subroutine OUTPUT will print when called after 4A, and stage 11 is called. IREAD = 1 and N4 = 1 after returning from stage 4A means an optimal result was read in from card type 11. The iteration limit and iteration count are set to be unequal, and stage 13A is called. At stage 5, IFØUN1 = 0 implies that no link was found at stage 4. After the search is ended for one budget constraint, the following debugging output can be printed: the next budget constraint (BUDGET) and network number

(MLIST), and the array of budget constraints (NSIZES).

Important variables not defined above are as follows:

MLIST - network number for the network to be worked on, which is also an index for arrays NSIZES and NITL.

MAXØD - maximum positive, odd element in array MLEVEL; if no positive, odd element exists, MAXØD = 1. MAXØD of one signals that the search is completed, and results should be printed.

BUDGET - current budget constraint, an INTEGER variable.

NCØST - total construction cost of all links (i,j) with MLEVEL $\geq$ 0 and with KFØRCE (i,j) $\geq$ 0 and KFØRCE(j,i) $\geq$ 0. A negative KFØRCE indicates the link is forced by the user to be excluded for this budget. KFØRCE is a symmetric matrix. Consequently, NCØST represents the total construction cost of links included (constructed) at the moment. Note that the link chosen at stage 4 which causes stage 7 to be entered is not constructed as a result of that choice. It is still a candidate for construction at stage 18 and at the next entry to stage 4 because its MLEVEL remains -1. Such a link will be constructed at stage 13 if not before. Links are only constructed at stages 8, 4A, 18, and 13.

N4 - binary indicator

0, a previously processed network was read at stage 4A; search should be resumed on this network. Do not print if subroutine ØUTPUT is called.

1, normal. Print during subroutine ØUTPUT.

IREAD - binary indicator

0, no network was read at stage 4A

1, a network was read at stage 4A. In combination with N4 = 1, IREAD = 1 means an optimal result was read at stage 4A.

IF8 - 0, after stage 6C, go to stage 7.

1, after stage 6C, go to stage 8 since 6D has already been performed.

MØW - iteration limit for current budget constraint.

NITE - number of iterations used on present budget; an iteration consists of MAXOD ≠ 1 at stage 12.

IFØUN1 - 0, no link was found at stage 4.

i, i > 0, a link was found at stage 4.

NØYES - 0, after stage 6, go to stage 8.

1, after stage 6, go to stage 6B.

Important arrays in MAIN are:

MLEVEL - see Section 4.8.5.

KLEVEL - columns 1-12 of KLEVEL correspond to the input of card type 6 with row i corresponding to the ith card read, if IFØR ≥ i. Row i, i > IFOR, is zero for all columns (1-16). Stage 4A sets up columns 13-16 of KLEVEL as follows: links to be ignored or not constructed for this budget constraint are listed by node numbers in columns 13 and 14, with the lower node number in column 13; links forced to be constructed by user for this budget constraint are listed in columns 15 and 16, with the lower node number in column 15.

MBEST - see Section 4.8.5.

NSIZES - contents of card type 3, budget constraints.

NITL - contents of card type 4, iteration limits.

IKNØWN - contents of card type 5, network numbers for networks previously processed.

TITLE - contents of card type 1, title.

F - for the FØRTRAN statements dealing with DØ loop "DØ 600", F is the matrix of minimum internodal distances with all NLINKS links constructed. F is an INTEGER array.

4.8.2 SETUP

SETUP is the first subroutine called by MAIN, and completes the execution of stage 1. SETUP reads card types 7-9, and initializes arrays MD and LD. Both have all values set to infinity (INFIN) except the diagonal. If a link (i,i) is read, the program prints an error message and stops. Note that in reading the links, the program assigns construction cost to lower half of LD, and link distance to upper half of LD no matter whether link is specified (i,j) or (j,i), i<j. MLEVEL is initialized to zero for the construction of the minimum spanning tree as required for subroutine SPAN. After calling SPAN, the minimum spanning tree has been found based on construction cost. MLEVEL = 3 for links in the spanning tree; and MLEVEL = 0 for links out of the spanning tree, immediately after returning from SPAN. The statements from "DØ 220" to "300 CØNTINUE" print which links form the minimum spanning tree.

The next set of code till "330 CØNTINUE" constructs the minimum spanning tree one link at a time, sets MLEVEL = 3 for links in minimum spanning tree, MLEVEL = -1 for links out of the minimum spanning tree, and sums the cost of constructing the tree.

The spanning tree's minimum internodal distances are printed if ISPAN = 1. LØWSUM is initialized according to stage 1.

IMPORTANT arrays not defined above, or redefined are:

F - after returning from subroutine SPAN, F is used as the array for printing the matrix showing which links are constructed for minimum spanning tree. F is an INTEGER array.

MD - matrix of minimum internodal distances for the links presently constructed. The links presently constructed have MLEVEL $\geq$ 0 and KFØRCE $\geq$ 0. Unless user is forcing links, KFØRCE = 0. The diagonal of MD is zero.

LD - upper half of LD is the link lengths. Lower half of LD is the link construction costs. Diagonal is zero.

FMT - user specified variable format from card type 8.

MLEVEL - array of integers giving the candidacy of links for construction. Just before calling SPAN, MLEVEL = 0. After completion of SPAN until the block beginning "DØ 330", MLEVEL = 3, if link is in minimum spanning tree and 0 otherwise. After "330 CØNTINUE," MLEVEL = 3 if link is in minimum spanning tree and -1 if not.

R - after calling SPAN, R is the first back node matrix of the minimum paths given the links presently constructed; R(i,i) = i. However, during execution of SPAN and until "300 CØNTINUE," R has another purpose as defined in SPAN. R is an INTEGER array.

The only important variable not defined in the input, or explained above, is LØWSUM, the current lowest sum of minimum internodal distances.

4.8.3 STAGE2

This subroutine performs stage 2 and a portion of stage 3 of the flowchart. Stage 2 of the flowchart is straightforward except for the computation of unconditional thresholds when NKNØWN = 0. The unconditional threshold, LTHRES, of a link is defined as follows:

0, for nonexistent (infinite length and costs) links

ℓ, where ℓ is the sum of minimum internodal distances over all pairs of nodes with all links constructed except the link whose unconditional threshold is being computed.

When NKNØWN = 0, array F is initialized to the link lengths, and B is set to B(i,j) = i for all (i,j). Both F and B are INTEGER arrays. Subroutine MINDIS computes the minimum internodal distances with all links constructed and the corresponding back node matrix. F is the matrix of minimum internodal distances and B is the corresponding back node matrix upon return from MINDIS. LSUM holds the corresponding sum of minimum internodal distances.

The procedure for computing unconditional thresholds is: for each finite link, delete it from the network, and compute its unconditional threshold. At this point MD holds the minimum distances with the spanning tree constructed, while F holds the minimum distances with all links constructed. The computation of removing a link from a network requires that MLEVEL correspond to the links actually constructed, taking into account user-forced links, which are indicated in KFØRCE. MLEVEL corresponds at this point to the minimum spanning tree and MD, but a link must be removed from the complete network. Consequently, MLEVEL is temporarily stored in MBEST, which has not been assigned any values yet. And MLEVEL is set to one for all links to indicate that all links are constructed in array F. Even for infinite links, MLEVEL = 1 with no consequences. For most of the program, however, infinite links have MLEVEL = -1.

Computation of unconditional thresholds is then executed for all links. If the link is infinite, it is assigned an unconditional threshold of zero. For finite links, F is loaded into ID, so that ID now has the internodal distances with all links constructed. LNKØUT deletes a link and leaves the corresponding minimum internodal distances in ID. At this point B has no meaning. LNKØUT places lists of nodes in M and N; MLIM is the number of nodes in M, and NLIM is the number of nodes in N. The set of all node pairs {M(I), N(I), I = 1 ... MLIM, J = 1 ... NLIM} lists all paths that could have been lengthened by the deletion of the link. If MLIM = 0 and NLIM = 0 (they will always both be 0 or both greater than 0), the link was useless with all links constructed and its unconditional threshold equals the sum of minimum distances with all links constructed. Therefore, such links are redundant. With MLIM > 0 and NLIM > 0, some paths may have been increased. If the link had zero length, just sum the new minimum internodal distances, for both M and N contain all nodes. If the link had positive length, the unconditional threshold is the sum of increases in internodal distances. Since all reverse paths are also lengthened for each link, {M(I), N(J)}, lengthened, the increase is twice LDIFF.

MLEVEL is restored to the minimum spanning tree after all unconditional thresholds are computed by setting MLEVEL = MBEST. Note that the storing of MLEVEL temporarily in MBEST now executes part of stage 3: set MBEST = MLEVEL, for all cases with NKNØWN = 0. The MAIN program sets MBEST = MLEVEL after stage 2 no matter what the value of NKNØWN.

Important arrays not previously defined in STAGE2 or redefined are:

LTHRES - unconditional thresholds. If NKNØWN > 0, card type 10 is read in here. Note the methods of storage as described at the outset of Section 4.8.

B - up to "DØ 100," this is the back node matrix for minimum paths with all links constructed. During and after "DØ 100," it has no meaning. B is an INTEGER array.

F - minimum internodal distances with every link constructed. F is an INTEGER array.

MBEST - 3, if link is in minimum spanning tree.

-1, if link is not in minimum spanning tree.

Undefined if NKNØWN ≠ 0.

MLEVEL - see detailed description above on coding of STAGE2.

ID - work array used in computing unconditional thresholds; see the detailed description above on the coding of STAGE2.

M - array produced by subroutine LNKØUT. If link (j,k) is deleted, M contains the nodes i, which could have used link (j,k) in the shortest path from i to k.

N - array produced by subroutine LNKØUT. If link (j,k) is deleted, N contains those nodes ℓ which could have used link (j,k) in shortest path from j to ℓ. The set of all (M(m), N(n)) gives all paths that could be lengthened by deleting link (j,k) from network.

Variables not defined before are:

NKNØWN - defined on card type 2 of input.

NL - defined in MAIN program.

LSUM - sum of minimum internodal distances with all links constructed.

INFIN - defined in MAIN.

N1 - defined in MAIN.

LDIFF - for positive, finite length links, the sum of increases in paths lengthened upon deleting a link; as the miminum paths are only computed in one direction, the increase in sum of minimum distances is twice LDIFF. For zero length links, LDIFF is the sum of minimum internodal distances with all links constructed except the link whose threshold is being computed.

MLIM - number of nodes in array M.

NLIM - number of nodes in array N; for meaning of MLIM = 0 or NLIM = 0, examine detailed description of coding of STAGE2 or see subroutine LNKØUT.

The debugging output printed in computing unconditional thresholds is:

1. Minimum internodal distances and corresponding back node matrix with all links constructed; actually printed by subroutine MINDIS.

2. Sum of minimum internodal distances with all links constructed.

3. Unconditional thresholds; note the storage procedure in array LTHRES as explained above to find the unconditional threshold of a given link.

4.8.4 STAJ4A

This subroutine does much of the work for two options: reading in previous results by NKNØWN > 0, and forcing links by IFØR > 0. Introductory explanations of those options are given in Section 4.4.

If NKNOWN = 0, the code dealing with reading in previous results is skipped; and N4 = 1, IREAD = 0. For NKNØWN > 0, the subroutine searches array IKNØWN (read from card type 5) to determine if there is input from previous jobs to be read. There is input to be read for this budget constraint if the network number (MLIST) for this budget constraint is on card type 5. Otherwise, skip to the coding on forcing links after setting IREAD = 0 and N4 = 1. If this budget constraint does have a network from previous jobs, read a series of cards of card type 11. If the series of cards is not for the current budget constraint, print an error message and stop.

If a network not known to be optimal was read, N4 = 0 meaning the search should be restarted where it was stopped in a previous job by iteration limit. The cards with MLEVEL array are read into array MBEST. Then, subroutine ØUTPUT is called with N4 = 0. Its purpose is to construct the appropriate new network. The current MLEVEL array tells what links are now constructed. The MLEVEL array read from cards into array MBEST tells what links are constructed in the network being read in. Calling ØUTPUT with N4 = 0 has the effect of constructing the links that should be constructed for the network read in; no printing is done. Upon return from ØUTPUT, MLEVEL is set to MBEST so that the MLEVEL array read from card type 11 is finally stored in MLEVEL. The best sum of minimum internodal distances and MBEST are read from card type 11. IREAD = 1, MAXOD = 1, and the subroutine transfers to the coding on forcing of links. N4 is still zero.

If a network known to be optimal was read, N4 = 1 on card type 11. The MBEST array of card type 11 is read into MBEST. IREAD = 1, MAXØD = 1, and the coding on forcing links is started. N4 is still one. If there is no forcing at all for this job, exit the subroutine. Otherwise, initialize columns 13-16 of KLEVEL.

Links that are specified to be forced are examined to see if they are to be forced for this budget constraint. This is accomplished in statements "IF(KLEVEL(I,3).EQ.100) GØ TØ 180" through "GØ TØ 220." For links forced to be constructed or excluded for this budget constraint, set their MLEVEL and MBEST to zero. Store information in KLEVEL columns 13-16 and array KFØRCE as given in array descriptions below. Construct or remove links from the network as is needed to permit forcing.

If there has been no forcing for this budget constraint, or if there has been forcing but card type 11 was read for this budget, do not call subroutine STG4AB. Otherwise, enter STG4AB, as the best sum of minimum distances must be recomputed and links need to be removed if the budget constraint is too restrictive.

The debugging output should be self-explanatory. Important array definitions are:

IKNØWN - contents of card type 5.

MBEST - when cards from card type 11 are read with N4 = 0, see the description of coding in this section. Otherwise, an entry of i, i > 0, means that the link is constructed in the best network found so far; an entry of i, i < 0, means that the link is not constructed in best network thus far; an entry of zero for link (j,k) and KFØRCE = 0 means link (j,k) is constructed in best network thus far; an entry of zero for link (j,k) and KFØRCE > 0 means link (j,k) is forced for this budget. When KFØRCE(j,k) > 0 the link is forced to be constructed. KFØRCE(j,k) < 0, the link is forced to be excluded. Note MBEST and MLEVEL on card type 11 truly reflect the link's being constructed or not constructed, as forcing of links does not affect MBEST and MLEVEL on the punched output.

MLEVEL - i > 0 means link is currently constructed.
i < 0 means link is currently unconstructed.
i = 0 for link (j,k) and KFØRCE(j,k) = 0 means link is currently constructed
i = 0 for link (j,k) and KFØRCE(j,k) ≠ 0 means link is forced for this budget constraint. If KFØRCE(j,k) > 0, (j,k) is currently constructed. KFØRCE < 0, implies that (j,k) is not currently constructed.

LTHRES - unconditional thresholds.

KLEVEL - columns 1-12: row i, i $\leq$ IFØR is the ith card of type 6; row i, i > IFØR is unused.
columns 13, 14: link user has forced to be excluded for this budget constraint. Column 13 contains lower node, and column 14 the higher node. All entries after the end of the list are zero.
columns 15, 16: link user has forced to be constructed for this budget constraint. Column 15 contains lower node, and column 16 the higher node. All entries after the end of the list are zero.

KFØRCE - 1 in entry (i,j) means link (i,j) has been forced by user to be constructed for this budget constraint.

-1 in entry (i,j) means link (i,j) has been forced by user to be excluded for this budget constraint.

0 in entry (i,j) means link (i,j) is not forced by user for this budget constraint.

The matrix is symmetric; diagonal is zero.

MD - minimum internodal distances for links presently constructed; diagonal is zero.

R - back node matrix for minimum paths corresponding to MD. R(i,i) = i. R is an INTEGER array.

LD - upper triangular matrix contains link distances. Lower triangular matrix contains link costs; diagonal is zero.

Important variables redefined or not defined before are:

N4 - binary indicator;

0, the network read in from card type 11 was not necessarily optimal; search should be resumed at point where it stopped. When subroutine ØUTPUT is called, print nothing, but construct and remove links as needed.

1, the network read in from card type 11 was proven optimal; when subroutine ØUTPUT is called, print normally.

On returning to MAIN with N4 = 0, MAIN completes initializations to restart search and then sets N4 = 1.

IREAD - binary indicator;

0, none of card type 11 read in STAJ4A.

1, a series of cards of card type 11 was read.

LØWSUM - lowest sum of minimum internodal distances found so far in the search.

NCØST - cost of constructing the network which is used to compute the minimum distances MD.

4.8.5 Arrays and Variables with Same Definitions Throughout Remaining Subroutines

There are two exceptions to this list. In subroutine STAJ18, array MLEVEL is defined differently for the duration of the subroutine. Array R has a different definition in subroutines CØNECT and SPAN which find the minimum spanning tree. See the description of subroutine CØNECT for the local definition of R in CØNECT and SPAN.

MD - matrix of current minimum internodal distances, given the links currently constructed; diagonal is zero.

R - first back node matrix giving routes corresponding to the minimum distance of MD. R(i,i) = i.

NCØST - construction cost of the network currently constructed, which is the construction cost of the network that produced MD. Note that the final link whose selection at stage 4 yields a "YES" at stage 6 is not constructed, although for the purpose of computing the lowest sum of minimum internodal distances it is considered constructed. Because that link is not constructed, it is not included in computing MD, R, or NCØST. Also, since it is not constructed, it could be selected and constructed at stage 18 or at stage 8, after stage 18 is over.

LD - matrix of link information. Upper triangular matrix gives link lengths. Lower triangular matrix gives link construction costs. Diagonal is zero.

LØWSUM - lowest sum of minimum internodal distances found yet under this budget constraint.

BUDGET - current budget constraint; an INTEGER variable.

MLEVEL - array of integers with one entry for each link; see Section 4.8 for method of locating and storing the value of MLEVEL corresponding to link (i,j). For any link with MLEVEL(k) = i:

i > 0 means the link is currently constructed, and therefore used in computing MD, R, and NCØST.

i < 0 means the link is currently not constructed, and therefore not used in computing MD, R, and NCØST.

i = 0, if user has forced no links, means the link is currently constructed and used for computing MD, R, and NCØST.

i = 0, if user has forced links, means the link is currently constructed, if user has requested it forced for this budget constraint. Program determines this condition by KFØRCE = 1 for the link. The link is currently unconstructed if user has requested it to be excluded for this budget constraint. Program determines this by KFØRCE = -1 for the link. The link is also currently constructed if user has not forced the link for this budget constraint; i.e. KFØRCE = 0.

MBEST - array of integers with one entry for each link; see Section 4.8 for the method of locating and storing the value of MBEST corresponding to link (i,j). For each entry the definitions are the same as MLEVEL except that MBEST indicates what links are constructed and unconstructed in the best network found for this budget constraint, whereas MLEVEL tells what links are constructed and unconstructed in current network. For example, an entry in MBEST of i > 0 means the link is constructed for best network found thus far. The links of best network are independent of computing MD, R, and NCØST until stage 13. However, LØWSUM is the sum of minimum internodal distances for network described by MBEST.

N1 - NØDES - 1

NL - NØDES(NØDES-1)/2

INFIN - see MAIN program

MAXØD - maximum, positive odd value in MLEVEL; if no value exists, MAXØD = 1.

4.8.6 STG4AB

This subroutine is called only from subroutine STAJ4A and performs stage 4AB. Its purpose is to complete preparations for the forcing of links. None of these initializations are needed if input of card type 11 was read at stage 4A; therefore, the subroutine is not called in that event.

After the subroutine is called, LØWSUM must be updated since the lowest sum of minimum distances for a budget constraint with forced links could be greater than the lowest sum of minimum distances for a previous, smaller budget constraint which had no links forced. Also, it is possible by forcing, that too many links are constructed. Consequently, the subroutine will, if necessary, remove nonforced links from the network one-at-a-time. The one removed is the nonforced link whose removal least increases the sum of minimum internodal distances. This procedure stops when the cost of the constructed network is low enough to permit at least one more link with MLEVEL = -1 to be constructed within the budget constraint.

The coding down to statement "10 INDEX = 0" updates LØWSUM. The coding from statement "10 INDEX = 0" until statement "IND = 0" checks that at least one more link with MLEVEL = -1 can be constructed within budget constraint. The "DO 60" finds the least useful nonforced link presently constructed, i.e. the link that increases the sum of minimum internodal distances least. Note that KFØRCE matrix has entries defined for all links, whether their MLEVEL = 0 or not. Consequently, of all links considered for removal, each link must have its MLEVEL > 0 and its KFØRCE = 0. Matrix F is used to compute the minimum distances with one link deleted. F is an INTEGER array. If MLIM = 0, the link under consideration was useless in that it wasn't used for any minimum path. Therefore, the first such link is chosen, if there is one. Otherwise, for each useful link, the increase in sum of minimum internodal distances upon removing the link is computed in variable KDIFF.

The program stops completely if there are so many links forced into construction that the budget constraint does not permit the construction of any link except for the forced links. The subroutine detects this after completing the "DØ 60" loop. The subroutine must still remove a link at that point. If MINN = LIMIT, then, there are no links that could be removed because all are forced. If the above situation does not happen, the link chosen for removal is then removed. Its MBEST and MLEVEL are set to -1, and its cost subtracted from NCOST.

To update LØWSUM, the subroutine need only add MINN to LØWSUM, for MINN contains the minimum increase in the sum of minimum internodal distances, the increase resulting from the link being removed. Then, the budget constraint is checked to determine if enough links have been removed. This check is again the block of coding from statement "10 INDEX = 0" to statement "IND = 0." Should more links need to be removed, repeat the process starting at statement "30 MINN - LIMIT."

Because LØWSUM has been updated, and perhaps increased because of forcing of links, all nonforced links with MLEVEL = 0 must be checked to insure that their unconditional thresholds are still greater than or equal to LØWSUM. For, among nonforced links, the only links that should have MLEVEL = 0 are those which have unconditional threshold greater than or equal to LØWSUM. For every nonforced link with MLEVEL = 0 and unconditional threshold less than LØWSUM, set its MLEVEL and MBEST to 3. This is the coding from "IND = 0" to "28 CØNTINUE." This procedure need only be performed once, just before returning to STAJ4A.

Important variable definitions not defined above are:

LIMIT - $(2^{31}-1)$

MINN - used in minimizing.

MLIM - defined by subroutine LNKØUT.

NLIM - defined by subroutine LNKØUT.

KDIFF - increase in sum of minimum internodal distances upon removing a link. KDIFF is used for each link it is possible to remove.

Important arrays not defined above are:

F - work array used in computing minimum distances upon removing a link from network. F is an INTEGER array.

B - contents in this subroutine are meaningless; B is an INTEGER array.

LTHRES - array of unconditional thresholds; storage of threshold for link (i,j) is the same as in arrays MLEVEL and MBEST.

M - defined by subroutine LNKØUT.

N - defined by subroutine LNKØUT.

The subroutine prints output, even if IØ = 0. Format statement number 25 and the output itself are self-explanatory. As debugging output, the subroutine prints the links as they are considered for removal and the increase in sum of minimum internodal distances that would result from removal of each link. After a link is removed, the subroutine prints NCØST, MLEVEL of the link and the updated LØWSUM.

4.8.7 STAGE4

This subroutine performs stage 4 of the flowchart; note that infinite length links are excluded. To find which link decreases the sum of minimum internodal distances most, the subroutine finds the greatest difference in the sums of minimum internodal distances before and after construction of each link. Recall the definitions of arrays M and N given in Section 4.2.1. The difference in sums of minimum internodal distances before and after constructing link (i,j) has the following terms:

1. difference in minimum distances between i and j;
2. differences in minimum distance between M(k) and j, where k ranges over all of array M;
3. differences in minimum distance between i and N(ℓ) where ℓ ranges over all of array N;
4. differences in minimum distance between M(k) and N(ℓ) where k and ℓ range over all node pairs with an M element and N element.

It is possible for M or N or both to be the empty set. The program must make calculations appropriately.

Note subroutine LINKIN's definitions of MLIM and NLIM:

MLIM - (-1), the link if constructed would not change any minimum distance.

i, i $\geq$ 0, number of elements in array M.

NLIM - (-1), the link if constructed would not change any minimum distance.

i, i $\geq$ 0, number of elements in array N.

The coding should be self-explanatory. Note that the difference is multiplied by two; although the difference computed takes in all node pairs, paths are only considered in one direction, the reverse path not being taken into account. Note also that even if the construction of the link (i,j) would change no minimum distances, the program still computes the difference in minimum distance between i and j and doubles this difference. The result is zero, of course, because the distance did not change.

Important new variables are:

IFØUN1 - if a link was selected, IFØUN1 is the lower node of the link chosen. If no link passed the criteria, IFØUN1 = 0.

IFØUN2 - if a link was selected, IFØUN2 is the higher node of the link chosen. If no link passed the criteria, IFØUN2 = 0.

IFØUN3 - index for MLEVEL and MBEST. The index locates the entry in those arrays corresponding to the link selected. An equivalent way to define IFØUN3 is NØDES(NØDES-1)/2 - NØDES + IFØUN2 - (NØDES - IFØUN1)(NØDES -1 - IFØUN1)/2 The program does not compute it this way, however. If no link was chosen, IFØUN3 = 0.

MX - maximum decrease in sum of minimum internodal distances. The decrease comes from constructing link (IFØUN1,IFØUN2). If no link was chosen MX = -1.

NSTRAN - BUDGET less NCØST; amount of budget still available for constructing links.

MDIFF - used to compute the difference in sum of minimum internodal distances for each link considered.

Important arrays are:

F - an INTEGER array used to compute minimum distances with a new link constructed.

B - contents are meaningless. B is an INTEGER array.

M - defined in subroutine LINKIN.

N - defined in subroutine LINKIN.

For debugging output, every link that is a possibility for selection at STAGE4 is printed along with the decrease in the sum of minimum internodal distances each link would produce if constructed. At the end of STAGE4, the link (IFØUN1, IFØUN2), MX, and IFØUN3 (link number) are printed.

4.8.8 <u>STAGE6(NØYES)</u>

This subroutine asks the question given in stage 6 of the flowchart. NØYES is a binary return variable telling whether the answer was NØ or YES. Of the links with MLEVEL = -1, the cost of the minimum construction cost link other than link (IFØUN1, IFØUN2) is identified. To exclude link (IFØUN1, IFØUN2), the index for array MLEVEL is forced not to be IFØUN3. If the minimum construction cost found above plus the cost of link (IFØUN1, IFØUN2) plus the cost of network presently constructed is less than or equal to the budget constraint, the answer to the flowchart stage 6 question is YES, NØYES = 0, and another link can be constructed. Otherwise, the answer is NØ, NØYES = 1 and no link at MLEVEL = -1 can be constructed in addition to (IFØUN1, IFØUN2), for not even the minimum cost link at MLEVEL = -1 could be added with (IFØUN1, IFØUN2).

Note that in stage 6B of the flowchart, condition three is the same as the question of stage 6. Therefore, subroutine STAJ6B also calls stage 6 subroutine to ask this question. IFØUN1, IFØUN2, and IFØUN3 are then redefined temporarily by STAJ6B in order to use stage 6.

All arrays are as defined earlier. Important variables are:

NØYES - binary return variable described above.

MN - variable used to find the minimum construction cost of a link with MLEVEL = -1 but other than link (IFØUN1, IFØUN2).

For debugging output, the value of NØYES is printed along with messages defining its meaning.

4.8.9 STAGE8

This subroutine performs the simple tasks of stage 8 of the flowchart. Note that in constructing the link, subroutine LINKIN is called; all minimum distances (array MD), and first back nodes of minimum paths (array R) are updated. R is an INTEGER array. All arrays and variables are as defined earlier.

4.8.10 CHKVAR

Subroutine CHKVAR performs stages 7, 9, and 10 of the flowchart. Also, some output is printed here to inform the user of theprogress of the search. Stage 7 is always executed. Upon entry of stage 7, there always are two networks, the best network up to this point in the search and a network to be compared with it. For the first budget constraint and the first entry of stage 7, the best network thus far is just the minimum cost spanning tree while the new network for consideration would be the spanning tree plus all links selected at stage 4. This would not be the case if a solution from cards were read at stage 4A, for the card solution would give the best network and the new network to which stage 4 would add links.

Another special case would occur on the first entry to stage 7 for a budget with forcing of links, but no card type 11. Then the best network would be the one as determined in stage 4AB and the new network for consideration is that network plus all links selected by stage 4. In all other cases the best network would be the one as defined by stages 9 and 10 the last time they were executed or by card type 11, if this was the first execution of stage 7 after reading at stage 4A. The new network to be considered would be the network generated by the search which uses stages 4, 16, 17, and 18 repeatedly.

The sum of minimum internodal distances for best network is stored in LØWSUM. The sum of minimum internodal distances for the new network under consideration is computed and stored in NEWSUM. Recall that the minimum distance array MD has not been updated for the construction of the last link (IFØUN1, IFØUN2) selected at stage 4. Therefore, the construction of (IFØUN1, IFØUN2) must be taken into account to find NEWSUM. That is accomplished by variable MX. If NEWSUM < LØWSUM, the answer to stage 7 question is YES. Otherwise, the old best network is better or at least as good, as far as the sum of minimum internodal distances, as the new network considered.

Some main output is printed at this point. Stage 7 always prints the links that constitute the new network being considered, the sum of minimum internodal distances for that network, LØWSUM, and messages indicating whether this network is the best found yet or that a previous network is the best. This concludes stage 7. If NEWSUM $\geq$ LØWSUM, the subroutine exits. Otherwise, stages 9 and 10 are executed in CHKVAR. Stages 9 and 10 are self-explanatory. The subroutine exits on completion of stage 10.

Note that additional printing is performed whenever CHKVAR is exited, if the user has MØUT = 1 on card type 7. In this case, MLEVEL array is printed at the time of exit. Consequently, if the new network is not better than all previous networks, MLEVEL corresponds to the end of stage 7. When the new network is better than the previous best, MLEVEL corresponds to the completion of stages 9 and 10. The format for printing the array is to print the entry number followed by the value of MLEVEL; 14 entries with their corresponding values are printed per line.

Important arrays are:

KFØRCE - square array with zero diagonal. If entry (i,j) equals:

1, link (i,j) must be constructed as user wishes for this budget constraint (it is forced into construction);

0, link (i,j) is not forced for this budget constraint;

-1, user has forced link (i,j) not to be constructed for this budget.

LVP1 - lower node numbers of the list of links of the new network being considered. IFØUN1 is not in the list.

LVP2 - higher node numbers of the list of links of the new network being considered. IFØUN2 is not in the list.

LTHRES - array of unconditional thresholds.

Important variables are:

NEWSUM - sum of minimum internodal distances for the new network being considered.

MX - variable passed from STAGE4. MX is the decrease in sum of minimum internodal distances resulting from link (IFØUN1, IFØUN2).

JL - number of elements in LVP1.

IFØUN1, IFØUN2, IFØUN3, and MX are as defined in subroutine STAGE4.

4.8.11 FMAXØD

This subroutine performs stage 11 of the flowchart. All arrays and variables are as defined above. Note the definition of MAXØD as given in stage 11 of flowchart and Section 4.2.2. MAXØD = 1 is the signal to stop the search and print results.

4.8.12 STAJ16

This subroutine performs stage 16 of the flowchart. The coding is straightforward. Note that to remove a link from the network, subroutine LNKØUT is called to update MD (minimum distances) and R (first back node matrix). R is an INTEGER array. All arrays and variables are as defined before. As debugging output, STAJ16 prints the value of NCØST and the contents of array MLEVEL.

4.8.13 STAJ17

This subroutine performs stage 17 of the flowchart. Of the links with MLEVEL = MAXØD, find the link which gives the least increase in sum of minimum internodal distances when removed from network. The coding to do this is more complicated than one might expect and takes up the first section of the subroutine down through statement "90 CØNTINUE."

For each link with MLEVEL = MAXØD, the link must be removed to determine the effect of its removal on the sum of minimum internodal distances. For each link with MLEVEL = MAXØD, the following process occurs. The minimum distances (MD) of network currently constructed are copied into array F, which is an INTEGER array, also. The link is then removed from the network, and F contains the minimum distances of the network without the link. To remove the link, subroutine LNKØUT prepares two arrays M and N whose Cartesian product gives all paths that might have been lengthened by removal of the link. Hence, to calculate the increase in sum of minimum distances on removing the link, the program sums the increases in minimum paths with the Cartesian product of M and N as the index set. Note that MLIM tells the number of elements in array M and NLIM the number of elements in array N. If the link distance of the link being considered is zero, the sum includes all paths in both directions. However, if the link distance is positive, the Cartesian product of M and N only includes paths in one direction; to get the reverse paths, the sum of increases must be doubled. The increase in sum of minimum internodal distances is then checked to see if it is the minimum increase of the links processed so far.

If the network contained a useless link, i.e. a link not needed for any minimum path, the minimizing process is stopped at the first such link found. That link is immediately selected as the link to be removed at stage 17. Such a link is detected by subroutine LNKØUT returning a value of MLIM = 0; the test is performed two statements after the call to subroutine LNKØUT. The rest of the coding of stage 17 is straightforward. Note that after the link to be removed is found, subroutine LNKØUT must be called again to actually remove the link and have this recorded in arrays MD and R.

Important arrays are:

F - integer work array used to store minimum distances of network with one link removed.

B - a dummy integer array needed for subroutine LNKØUT.

M and N are defined by subroutine LNKØUT.

Important variables are:

MN - current minimum increase in sum of minimum internodal distances.

KØUT - lower node of the link finally selected to be removed at stage 17.

LØUT - higher node of the link finally selected to be removed at stage 17.

KINDEX - index of MLEVEL array corresponding to link (KØUT, LØUT).

IDIFF - used to compute the increase in sum of minimum internodal distances upon removing a link.

MLIM and NLIM are defined by subroutine LNKØUT.

For debugging output, STAJ17 prints the increase in the sum of minimum internodal distances if link (i,j) were removed, for each link (i,j) with MLEVEL = MAXØD. Also a summary of the results of stage 17 are shown by printing the link selected (the link "LEAST HELPFUL"), the link's updated MLEVEL, new value of NCØST, and the increase in the sum of minimum internodal distances that removal of the link gave.

4.8.14 STAJ18

This subroutine performs stage 18 of the flowchart. Its main purpose is to compute conditional thresholds and construct links according to the conditional thresholds. For all links with MLEVEL $\neq$ -1, the conditional threshold is zero. Also, as indicated on the flowchart, the subroutine may exit because of the construction cost condition. This exit could occur before the conditional thresholds of all links with MLEVEL = -1 are computed. For any link whose conditional threshold is not computed, its conditional threshold is zero.

The purpose of stage 18 is to take the network as constructed after stage 17 and add links to the network. The selection of additional links is refined by the use of conditional thresholds; see statements 1 and 2 in stage 18 of flowchart. However, upon exit from stage 18, the network construction cost must be small enough so that at least one unconstructed link at MLEVEL = -1 may be constructed without exceeding the budget constraint. This is the purpose of statement 3 in stage 18 of the flowchart.

To compute conditional thresholds, time is saved by constructing a working network in another array. This network consists of the network existing at the end of stage 17 plus all finite length links with MLEVEL = -1 or MLEVEL < -MAXØD. The loop "DØ 50" finds all links that must be added. This DØ loop sets up a list of the links with the lower nodes stored in array LVP1 and the upper nodes in array LVP2. Variable JL gives the total number of such links.

Constructing a working network in addition to the actual network does complicate the computation. Array MSAVE is used to hold the MLEVEL values associated with the actual network. Array MLEVEL is then used to hold the "MLEVEL" values associated with the temporary working network to be constructed. This is very important; subroutine LNKØUT in removing a link from a network uses array MLEVEL to determine whether a link is constructed or not. Since computing a conditional threshold requires the removal of a link from the working network, creating MSAVE and MLEVEL arrays was essential. Note that subroutine LINKIN, to add a link to the network, does not need information concerning what links are presently constructed.

To construct the working network, two methods could be used: the Floyd-Murchland one-pass matrix method could construct it without using the minimum distances of the smaller actual network; or the Murchland algorithm could be used to add the links to the actual existing network one link at a time. By rough estimates, the Murchland link addition algorithm would be more efficient when the number of links to be added is less than 4(NØDES). However, if more links are to be added, the Floyd-Murchland one-pass method should be the more efficient. The appropriate method is chosen based on this criterion.

The coding to add a link one at a time is the loop "DØ 80", which is straightforward. To set up for the Floyd-Murchland algorithm is more complicated. Both methods use array F for the minimum distances for the working network and array B for the corresponding first back node matrix. Both F and B are INTEGER arrays.

The coding to prepare for the Floyd-Murchland method begins with statement "90 INDEX = 0" and ends with statement "115 F(L,K) = LD(K,L)." The purpose of loop "DØ 110" is simply to define array B as B(i,j) = i, for all (i,j), i $\neq$ j. The diagonal is not redefined; it should be B(i,i) = i. The second purpose of loop "DØ 110" is to define array F as

F(i,j) = link distance of (i,j), if (i,j) is constructed in actual network;
∞, if (i,j) is not constructed in actual network;
0, for diagonal.

Note that at this point MLEVEL and MSAVE have the same contents, so that both correspond to the actual network. Loop "DØ 115" then updates array F so that for all links (i,j) that are to be added to the actual network, F(i,j) = link distances of (i,j) and F(j,i) = link distance of (i,j). Then subroutine MINDIS executes the Floyd-Murchland algorithm.

At the end of either procedure MSAVE equals the network's "MLEVEL" at the end of stage 17, while array MLEVEL differs from array MSAVE only for the links added to that network to make the working network. For those added links MLEVEL = 1. After the procedure selected for the computation is completed, the sum of minimum internodal distances for the working network is computed and stored in variable LØBF.

The remainder of the subrotuine executes statements 1, 2, and 3 of the flowchart with loop "DØ 380." For finite links with MSAVE = -1 (MLEVEL for actual network), the conditional threshold of the link is the sum of minimum internodal distances of the working network without that link. Hence, to execute statement 1 of stage 18 of the flowchart, remove a link from the network, find the resulting increase in sum of minimum internodal distances, and add this increase to LØBF. The coding from statement "DØ 150 K = 1, NØDES" to statement number 165 executes statement 1. The matrix of minimum distances for the working network without the link whose threshold is being computed is located in array ID. Array B is meaningless, but necessary for subroutine LNKØUT. To compute the increase in the sum of minimum distances caused by the removal of the link, the same principle is used as in subroutine STAJ17, the increase being the sum of individual increases in path distances. Arrays M and N contain the essential information from subroutine LNKØUT. The Cartesian product of M and N gives all possible node pairs whose minimum paths could have lengthened. Hence, the sum is over (M x N). Loop "DØ 160" computes the sum. For zero length links, no additional computation is needed; but for positive length links, the sum must be doubled. Note that some links might be useless for the working network, and no minimum distances would change. The program detects this by MLIM = 0 after removing the link by subroutine LNKØUT. Then, the coding to compute the sum of increases is skipped.

Statement 2 of stage 18 is accomplished by one FØRTRAN statement. To execute statement 3 is the purpose of the remainder of the coding through the end of loop "DØ 380." Suppose the program is working on link (i,j); loop "DØ 360" finds the link (k,ℓ), other than link (i,j), which is the minimum cost link of those links with MSAVE = -1. (MSAVE is the MLEVEL for actual network). If this cost, plus the cost of link (i,j), plus the cost of actual network is greater than the budget constraint, the subroutine is completed; then MLEVEL array is restored by setting it equal to array MSAVE and the subroutine exits. Otherwise, link (i,j) is added to the actual network, NCØST is updated, and the MSAVE of link (i,j) is set to (MAXØD + 1), which means setting MLEVEL of actual network. This completes statement 3 of stage 18. Of course, it is possible that after all links with MLEVEL = -1 are tested by statements 1, 2, and 3, there still is budget for construction of more links. In that event, loop "DØ 380" terminates if it exhausts all links; MLEVEL is restored by setting it equal to MSAVE, and the subroutine exits.

Important arrays not defined above or in previous subroutine descriptions are:

CT - array of conditional thresholds; CT is an INTEGER array.

KFØRCE - square array with zero diagonal. If entry (i,j) is

1, user has forced link (i,j) into construction for this budget constraint.

0, link is not forced;

-1, user has forced link (i,j) to be excluded for this budget constraint.

Important variables not defined above are:

MM - -(MAXØD + 1).

MINN - used to find the minimum cost link with MSAVE = -1 as described in statement 3 of stage 18.

Debugging output for this subroutine is quite extensive. First, the links to be added to the actual network to obtain the working network are listed 10 per line under the heading "LINKS TØ BE PIVØTED IN TØ COMPUTE CØNDITIØNAL THRESHØLDS." Next the sum of minimum internodal distances of the working network is printed with label, "SUM ØF MINIMUM INTERNØDAL DISTANCES WITH APPRØPRIATE LINKS CØNSTRUCTED FØR CØMPUTING CØNDITIØNAL THRESHØLDS." The minimum distance matrix of the working network followed by first back node matrix is printed under labels "MATRIX MIN DIST FØR CØND THRESHØLDS" and "MATRIX BACK NØDE FØR CØND THRESHØLDS." These matrices are printed a row at a time. For each link (i,j) for which a conditional threshold is calculated, the following is printed:

1. "CØNSIDERING LINK" (i,j) "FØR ENTRY AT STAGE 18."
2. cost of link (i,j)
3. current value of NCØST (cost of actual network)
4. cost of link (k,ℓ) which is minimum cost link of statement 3 of stage 18. The label for that cost is "MIN CØST ØF ADDING ANØTHER LINK."

A summary is printed at the end of stage 18:

1. value of NCØST
2. value of budget constraint
3. cost of the last link (k,ℓ) found by statement 3 of stage 18. The label in output is misleading, "CØST ØF LEAST CØST LINK AT MLEVEL ØF -1." Actually the cost of last link whose conditional threshold was computed could be smaller. This entry could be ***** if that cost is infinite.
4. conditional thresholds printed 10 per line in the order in array CT. All links have a conditional threshold, though many are zero.
5. MLEVEL array as passed on in program. (MLEVEL = MSAVE.)

4.8.15 STG13A

This subroutine is the counterpart to STAJ4A. It would be useful to study the two subroutines together. STG13A undoes the special manipulations for forcing links, and punches card type 11 output for a subsequent job.

The coding to undo the forcing of links is in loop "DØ 20." Recall that columns 13 and 14 of matrix KLEVEL list links which the user has forced not to be constructed for this budget constraint. Consequently, one need only run down these lists to undo special considerations for forced links. Note that at the ends of the lists, the remainder of columns 13-16 of KLEVEL are zero. The number of links listed in columns 13 and 14 is counted by the subroutine with variable NØUT (number forced out of network) while the number of links in columns 15 and 16 is counted in NIN (number forced into network). For all forced links (i,j), KFØRCE (i,j) and KFØRCE (j,i) are set to zero by the subroutine. Hence, by the end of loop "DØ 20," all of array KFØRCE is zero. For all links forced by the user to be excluded for this budget, the DØ loop sets their MLEVEL and MBEST to -1. For all links forced by the user to be constructed at this budget, the DØ loop sets their MLEVEL and MBEST to 3.

Also, the program punches output with statement 100 through to the end of the subroutine. This output forms card type 11 input for a subsequent job; see description of the input card type for the format. Note that variable IØPT is the output name for the variable N4 to be punched by the program on card type 11. If NITE ≠ MØW, the optimal solution has been found, since the program stopped before the iteration limit. If NITE = MØW, the solution may be suboptimal, for the program was stopped by the iteration limit MØW. Note that for solutions known to be optimal, array MBEST is printed as well as punched. In the case of a solution halted by iteration limit, MLEVEL array, LØWSUM, and MBEST array are printed as well as punched.

Important arrays are:

KLEVEL - same as in STAJ4A subroutine.

KFØRCE - same definition as in STAJ4A upon entry to STG13A; upon exit of STG13A, the array is zero.

The definition of MBEST and MLEVEL are as before. Note though that no forced link has an MLEVEL or MBEST of 0, but rather an MLEVEL and MBEST reflecting its stage of construction in current network and in the best network.

Important variables are:

NØUT - number of links forced by user to be excluded for this budget.

NIN - number of links forced by user to be constructed for this budget.

IFØR - defined by user on card type 2.

IPUNCH - defined by user on card type 2.

NITE - number of iterations used in search of solution for this budget constraint.

IØPT - binary indicator:

0, the network solution may be suboptimal

1, the network found is the optimal solution

BUDGET - current budget constraint, an INTEGER variable.

The only debugging output comes immediately after the coding on forcing of links. If IFØR = 0, no debugging output will occur. A list of links (if any) forced by user to be excluded for this budget is printed with the label "LINKS FØRCED TØ BE IGNØRED," and a list of links (if any) forced by user to be in network for this budget is printed with label "LINKS FØRCED TØ CØNSTRUCTIØN." The MLEVEL and MBEST arrays are printed followed by the message "ARRAY KFØRCE SHØULD NØW BE ZERØ."

4.8.16 ØUTPUT

ØUTPUT performs stage 13 of the flowchart, most of which is straightforward. First, the subroutine constructs the best network found. This is accomplished in loop "DØ 30." Since the network constructed upon entry of the subroutine may not be the best network found, links must be removed from or added to the current network to obtain the best network. A link must be removed if the link's MLEVEL $\geq$ 0 while its MBEST < 0. A link must be added if the link's MLEVEL < 0 while its MBEST $\geq$ 0. Whenever such a link is removed from or added to the current network in building the best network, the link's MLEVEL is immediately updated by setting it equal to its MBEST. This is necessary since MLEVEL array must correspond to the network as constructed in order to compute minimum distances correctly when removing a link.

Immediately after completing the above procedures, the subroutine checks the binary indicator N4. If N4 = 0, subroutine ØUTPUT was called by STAJ4A when card type 11 was read. The entry to ØUTPUT was to prepare the program to start a search which had been halted by user specified iteration limit in an earlier job. The removal and addition of links is to convert the current network to the network read at STAJ4A, read from MLEVEL cards of card type 11 (read into array MBEST). Since this is now complete, subroutine ØUTPUT exits, and control returns to subroutine STAJ4A; see description of STAJ4A. However, if N4 = 1, the program knows that subroutine ØUTPUT is actually performing stage 13 of the flowchart rather than part of stage 4A. The rest of the subroutine is executed only when N4 = 1.

The program next recomputes LØWSUM. This is actually unnecessary, except in the case where an optimal result was read from card type 11. Note that just as in STG13A, subroutine ØUTPUT knows that the best network found is optimal when NITE < MØW; i.e., when the program completed the search before the user specified iteration limit was reached. When NITE = MØW, the search was halted by iteration limit; hence, the best network found could still be suboptimal. Messages stating whether network is guaranteed to be optimal or not are printed in the output. A thorough description of the output printed is given in section 4.6.

The following array definitions may be helpful:

TITLE - title read from card type 1 in MAIN.

KLEVEL - defined in subroutine STAJ4A.

ID - work array used to store the matrix showing which links are constructed. This matrix is prepared for printing. An entry (i,j) of ***** in the output (arising from trying to print variable INFIN in I5 format) means link (i,j) is not constructed. A numerical entry for (i,j) in the output gives the link length of the constructed link (i,j).

The following variable definitions are used:

N4 - binary indicator:

0, ØUTPUT was called from STAJ4A;

1, ØUTPUT is to perform stage 13 of the flowchart.

NITE - number of iterations required in the search.

MØW - user specified iteration limit on this budget constraint. If NITE < MØW, the network found is optimal. If NITE = MØW, the user specified iteration limit halted the search prematurely, and results could be suboptimal.

IFØR - defined by user on card type 2.

NIN - defined in STG13A.

NØUT - defined in STG13A. If NIN = NØUT = 0, there is no forcing of links for this budget constraint even though IFØR > 0. Therefore, the warning message about forcing links is not printed.

RØWSUM - INTEGER variable used to compute the row sums of minimum distance matrix.

NØDBACK - defined by user:

0, do not print back node matrix;

1, do print the first back node matrix.

4.8.17 RESET

Subroutine RESET performs stage 15 of the flowchart; the coding is straightforward. As intermediate output, arrays MLEVEL and MBEST are printed. All arrays and variables are as defined earlier.

4.8.18 CØNECT(M1,M2)

Subroutine CØNECT is called only from subroutine SPAN, which is the main coding for finding the minimum spanning tree. Subroutine CØNECT(M1,M2) is called to update the list of components used in selecting link (M1,M2) for the minimum spanning tree. To do this, it finds component A of M1 and component B of M2 and forms the new component (A U B). In that sense, the CØNECT connects nodes M1 and M2.

At the time CØNECT is used, no minimum paths have been computed; thus, MD and R are not yet used. To save storage, CØNECT uses array R to store the components. CØNECT is called many times by subroutine SPAN which itself is called by subroutine SETUP. After SPAN has found the minimum spanning tree, and after SPAN has returned control to subroutine SETUP, the construction (computation of minimum paths) of the minimum spanning tree found by SPAN begins. At that point, R becomes the first back node matrix for the minimum paths. The definition of R as the first back node matrix for minimum paths remains throughout the remainder of stage 1 and the rest of the program.

However, during subroutines SPAN and CØNECT, R has the definition given below. R is an INTEGER array. Suppose there are ℓ components, $2 \leq \ell \leq$ NØDES. For $i \leq \ell$, row i of R lists the nodes of component i; R(i, NØDES) = number of nodes in component i. For $1 \leq j <$ NØDES,

R(i,j) = node in component i, if $j \leq$ R(i, NØDES);

0, if j > R(i,NØDES).

For every row i, i > the number of components ℓ, the entire row is zero.

The above definition holds only if there is more than one component. Eventually, Sollin's algorithm to find the minimum spanning tree reduces the number of components to two, and CØNECT is called to connect the two nodes that join the components. When CØNECT is then called in this situation, the first two rows of R will only contain the two components. CØNECT detects that only one component containing all NØDES nodes will remain. R is unchanged except that CØNECT now sets R(1, NØDES) = NØDES, signaling that all nodes are connected. See Figure 4.2 for a flowchart of the subroutine.

Upon entry to subroutine CØNECT, the subroutine first determines if there is only one component remaining; if R(1, NØDES) = NØDES, the subroutine exits immediately, for that is the case; see stage 0. Stages 1 - 3 of the flowchart are performed in loop "DØ 20." When the component of M1 is found, the component's number is stored in LSAVE. That is, M1 is in the component numbered LSAVE. Loop "DØ 50" performs stages 4 - 6. Note that for stage 5, the question asked is: does M2 = R(k,i), where i takes all values $1 \leq i \leq$ R(K,NØDES) because R(K,NØDES) = number of nodes in component K. Stage 2 is handled similarly. Note that KSAVE stores the number of the component which actually contains M2.

The question of stage 7 is: R(LSAVE,NØDES) + R(KSAVE,NØDES) $\geq$ NØDES? Note that R(LSAVE,NØDES) + R(KSAVE,NØDES) $\leq$ NØDES, always when LSAVE $\neq$ KSAVE, because the combination of the two components cannot give more than the total number of nodes. Stage 8 is the statement R(1,NØDES) = NØDES, for this is the marker that the subroutine sets which is checked in stage 0. Stages 9 and 10 are performed in the statements from "80 NØLD = MINØ(LSAVE,KSAVE)" to "R(NØLD,NØDES) = ISTØP."

Figure 4.2 - Flowchart for Subroutine CØNECT

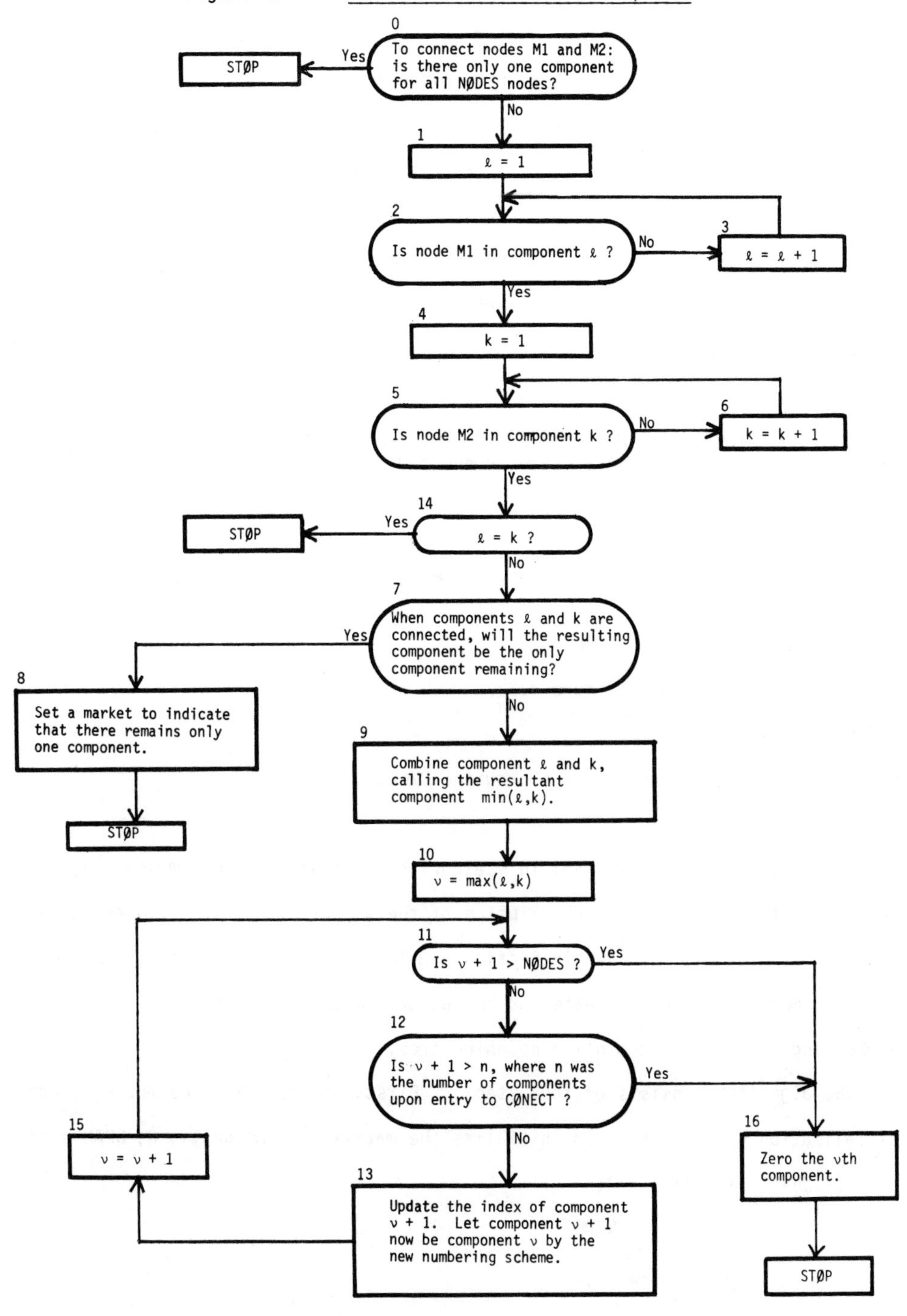

Stages 11 - 13 and 15 are performed in loop "DØ 100." Stage 13 actually amounts to transferring row (ν + 1) of R into row ν. The question of stage 12 is: R(ν + 1,NØDES) = 0? Recall that if originally there were no components, row i of R, for i > n, is zero. If there were n components at the entry of CØNECT and if the exit of CØNECT came from a YES to questions of stage 11 or 12, the subroutine zeroes row n of R; see stage 16.

For subroutines CØNECT and SPAN, R is as defined above. In all other subroutines, R is the first back node matrix for minimum distances of MD array. Important variables are:

M1 - a node

M2 - a second node, M2 $\neq$ M1. Link (M1, M2) has been selected for the minimum spanning tree; therefore, connect M1 and M2 and update the components.

LSAVE - node M1 is in the component numbered LSAVE; that is, M1 is in row LSAVE of R array.

KSAVE - node M2 is in the component numbered KSAVE. That is, M2 is in row KSAVE of R array.

NU - maximum (LSAVE,KSAVE)

NV - ν of flowchart, Figure 4.2.

NVP - ν + 1

There is no debugging output for subroutine CØNECT.

4.8.19 SPAN

Subroutine SPAN performs Sollin's algorithm to find the minimum spanning tree; see also Section 4.2.1 for a description of the algorithm itself. In addition, review the definition of R given in Section 4.8.18: R is a matrix that lists the nodes of each component which is created as the algorithm connects nodes and sets of nodes by choosing links for the minimum spanning tree.

The algorithm consists of four steps. First, however, SPAN performs several initialization procedures. To initialize the matrix of components, R, SPAN sets

$$R(i,j) = \begin{cases} i, & \text{if } j = 1; \\ 1, & \text{if } j = \text{NØDES}; \\ 0, & \text{otherwise.} \end{cases}$$

Sollin's algorithm requires that no two links have the same link cost. However, the input construction costs are integers, and some links may have the same cost. Consequently, SPAN sets up a new array DISTNZ in DØUBLE PRECISIØN as a matrix of perturbed link costs. Each link cost has a unique quantity r added to it, where $0 < r < 1$. Recall that LD is defined as:

$$LD(i,j) = \begin{cases} \text{link distance, if } i < j; \\ \text{link construction cost, if } i > j; \\ 0, \text{ if } i = j. \end{cases}$$

Let ε be a small number much less than 1. Define the perturbed matrix of link costs as

$$DISTNZ(i,j) = LD(i,j) + \left[(i - 1)(NØDES) + j\right]\varepsilon$$

If f(x) finds the greatest integer $\leq$ x, then ε must be chosen such that LD(i,j) = f(DISTNZ(i,j)). (Both the construction costs and link distances are perturbed to create DISTNZ, although only the construction costs are used. Let n be the maximum value that can be read as input for variable NØDES. Then, ε must be chosen such that $\varepsilon < 1/n^2$. Since n is currently 50, ε was set at 0.0003. Accordingly, the maximum perturbation is 0.7500; the minimum is 0.0003. Note that in the labelled CØMMØN block MATRIX, DISTNZ as a DØUBLE PRECISIØN array uses the same core storage as integer arrays F and B use in other subroutines. DISTNZ is destroyed after exiting subroutine SPAN.

The remainder of this description relies on the algorithm outline given in Section 4.2.2. Step 1 is performed in loop "DØ 50." Suppose for a given node i, with n = NØDES, the subroutine is searching for the node j such that

$$\text{cost}\ (i,j) = \underset{1 \leq k \leq n}{\text{minimum}}\left(\text{cost}\ (i,k)\right)$$

Variable AMX stores the minimum cost up to node k in the minimizing procedure. NEREST holds the node which gave that minimum cost. When the minimizing procedure is finished, link (i,NEREST) is selected for the minimum spanning tree. To indicate this, its MLEVEL is set to 3 and CØNECT is called to update the components.

Step 2 is performed by statement, "60 IF(R(1,NØDES).EQ.NØDES) RETURN." Note the description of R as given in description of subroutine CØNECT.

All of step 3 is performed by the loop "DØ 100," except that subroutine CØNECT has already numbered the components before entering the loop. Let ℓ be the total number of components, and k be any individual component. The statement, "IF(R(I,NØDES).EQ.0) GØ TØ 110" checks that step 3 has been completed for each of the ℓ components. For each component i, step 3 finds:

$$\underset{\substack{1\leq k\leq \ell \\ k\neq i}}{\text{minimum}}\left\{\underset{m'\varepsilon\, i}{\text{minimum}}\left\{\underset{n'\varepsilon\, k}{\text{minimum}}\left(\text{cost}(m',n')\right)\right\}\right\}$$

This is calculated by loop "DØ 90." The nested loops "DØ 80" and "DØ 70" perform the two innermost nested minimizations of the above calculation. In the coding, variable NØD1 is m' and NØD2 is n'. Variable BMX holds the minimum cost of the two innermost minimizations as they proceed. Variables NØD3 and NØD4 record the link (NØD3, NØD4) for which BMX was found. Variables AMX and BMX are DØUBLE PRECISIØN like DISTNZ. Variable AMX is used to hold the minimum cost up to the current point in the search in the outermost minimization. NØD5 and NØD6 record that AMX occurred for link (NØD5, NØD6). At the end of the calculation for each component i, the link found is stored in NCØN(i,1), NCØN(i,2). The value of ℓ is stored in variable NCØNLM upon exit of loop "DØ 100," so that the number of entries in the NØDES by 2 array NCØN is known. Step 3 actually says to find for each component i, the component j such that the above three minimizations are achieved. Actually, SPAN also stores node NØD5 of component i and NØD6 of component j which give the link cost sought. When CØNECT is called in step 4, the appropriate components i and j are found through NØD5 and NØD6.

Step 4 is performed by loop "DØ 120" and statement, "60 IF (R(1,NØDES).EQ.NØDES) RETURN." To revise the components, subroutine CØNECT is called for each link listed in array NCØN. To indicate that a link has been selected for minimum spanning tree, its MLEVEL is set to 3. Statement number 60, above, determines whether there is only one component remaining.

Important arrays are:

R - an INTEGER array defined in subroutine CØNECT. R(1,NØDES) equals NØDES means the minimum spanning tree has been found. In all subroutines other than SPAN and CØNECT, R is the first back node matrix for minimum paths for the network currently constructed.

DISTNZ - a DØUBLE PRECISIØN array defined in text above.

LD - as defined throughout program.

MLEVEL - defined in stage 1 of the flowchart.

NCØN - a NØDES by 2 array. Let ℓ be the number of components at step 3 of the algorithm. For $1 \leq i \leq \ell$, the link chosen by step 3 for component i is (NCØN(i,1), NCØN(i,2)). For $i > \ell$, the contents have no meaning.

The important variables are:

PTTHRE - value ε explained in the text above, currently 0.0003. PTTHRE is a DØUBLE PRECISIØN variable.

INFIN - integer used to represent infinity throughout the program.

DINFIN - DØUBLE PRECISIØN value of INFIN.

AMX - DØUBLE PRECISIØN variable defined in text used in two different minimization processes in step 1 and step 3.

NEREST - defined in text.

BMX - a DØUBLE PRECISIØN variable defined in text.

NØD1 - defined in text.

NØD2 - defined in text.

NØD3 - defined in text.

NØD4 - defined in text.

NØD5 - defined in text.

NØD6 - defined in text.

NCØNLM - total number of components during stage 3; i.e., the number of links listed in array NCØN.

As debugging output, matrix DISTNZ is printed one row at a time.

4.8.20 MINDIS

Given a set of links forming a network and the link lengths or distances of these links, find the minimum path distance between nodes i and k for each pair of nodes. Subroutine MINDIS solves this problem using the Floyd-Murchland algorithm. For a general description of the algorithm and related details, see Section 4.2.1 . The coding of the algorithm follows almost exactly the statement of the algorithm in Section 4.2.1.

Suppose not all of the possible links are to be constructed for the network whose minimum internodal distances are to be computed. Accordingly, define array F before calling subroutine MINDIS as follows:

F(i,k) = link distance of link (i,k) if i ≠ k, and if link (i,k) is to be constructed in the network;

∞, if i ≠ k and if link (i,k) is not to be constructed in the network;

0, if i = k.

Define array B before calling MINDIS as follows:

B(i,k) = i, for all (i,k).

Both F and B are INTEGER arrays. Arrays F and B used in subroutine MINDIS correspond exactly to the F and B matrices used in the description of the algorithm. In subroutine MINDIS, NØDES plays the role of n in the description of the algorithm.

Important arrays are:

F - an INTEGER matrix. For definition upon entry in MINDIS, see the above description. During MINDIS, a work array used to compute and store the minimum internodal distances.

B - an INTEGER matrix. For definition upon entry to MINDIS, see the above description. During MINDIS, a work array used to compute and store the first back nodes of minimum paths; i.e., at the conclusion of MINDIS,

B(i,k) = first back node on minimum path from i to k, if i ≠ k.

i, if i = k;

Important variables are:

N1 - NØDES - 1.

INFIN - integer representing ∞.

A - an INTEGER variable whose definition is clear in the coding.

As debugging output, MINDIS prints the label "SUBROUTINE MINDIS," the minimum distance matrix, printed a row at a time, and the first back node matrix corresponding to the minimum distances, printed a row at a time.

4.8.21 <u>LNKØUT(JP,KP,DP,PP,JPRINT)</u>

LNKØUT recomputes the minimum internodal distances of a network with link (JP,KP) removed from the network. For a definition of the problem and the statement of the algorithm used to recompute the minimum distances, see Section 4.2.1. The description of LNKØUT depends heavily on this statement of the algorithm, and most of the variables in the coding are identical to the symbols used in the algorithm description. The algorithm needs to know what subset of links are in the network as constructed; arrays MLEVEL and KFØRCE hold this essential information.

Assume link (JP,KP) is to be removed from the network. Matrix DP is the matrix of minimum internodal distances before removing (JP,KP) when the subroutine is entered. Matrix PP is the corresponding first back node matrix. Upon exit of the subroutine, DP and PP are the revised matrices of minimum internodal distances and first back nodes of minimum paths. Both DP and PP are INTEGER arrays. Variable JPRINT is unimportant to the algorithm, and therefore is defined later.

To perform step 1 of the algorithm, determine ℓ, the link length of link (JP,KP), which is located in the upper half of matrix LD. A few lines of coding insure that the upper half of matrix LD is used:

$$\text{SSTAR} = \ell = \begin{cases} \text{LD(JP,KP)}, & \text{if JP} < \text{KP}; \\ \text{LD(KP,JP)}, & \text{if JP} > \text{KP}. \end{cases}$$

SSTAR is an INTEGER variable. Then a check is made for SSTAR = ∞ or DP(JP,KP) < SSTAR. If either is true, the subroutine exits after setting NLIM = 0 and MLIM = 0, completing step 1.

Next consider step 2. Arrays M and N correspond to sets M and N in the description of the algorithm. Note as stated in step 2, if link distance of (JP,KP) = 0, then M = N and every node is in both M and N. However, if the link distance of (JP,KP) > 0, then (M $\cap$ N) is the empty set. Step 2 uses this result. If SSTAR = 0, the subroutine transfers to a more efficient set of code described below.

Assume SSTAR > 0 for the following description of steps 2, 3 and 4. Then, step 2 is performed in loop "DØ 20." MLIM gives the number of nodes in array M, and NLIM the number of nodes in array N. Since SSTAR > 0, (M $\cap$ N) must be a null set. Hence, if

$$\text{DP(I,KP)} = \text{DP(I,JP)} + \text{DP(JP,KP)},$$

there is no need to check for

$$\text{DP(JP,I)} = \text{DP(JP,KP)} + \text{DP(KP,I)},$$

for this cannot be true.

Step 3 is performed in loop "DØ 50." To check whether a link (i,ℓ) is to be constructed in the network or not, two statements are used.

```
    IF(MLEVEL(INDEX))  36, 26, 30
26  IF(KFØRCE(I,L) .LT. 0)  GØ TØ 36
```

Variable INDEX has been assigned the proper value so that MLEVEL(INDEX) is the MLEVEL corresponding to link (i,ℓ). If for link (i,ℓ), its MLEVEL < 0, the link is not constructed; if its MLEVEL > 0, the link is constructed in the network; if its MLEVEL = 0 and KFØRCE(i,ℓ) < 0, the link is not constructed in the network; if its MLEVEL = 0 and KFØRCE(i,ℓ) > 0, the link is constructed in the network; and if its MLEVEL = 0, and KFØRCE(i,ℓ) = 0, the link is constructed in the network. The statements "WRITE(6,34)" and "WRITE(6,35)" in loop "DØ 50" write a lengthy error message and the contents of arrays M and N along with JP,KP. These write statements are executed only if a severe error has been uncovered in the program. The error is that the link length of (JP,KP) > 0 and $(M \cap N)$ is not the null set.

Note that the MLEVEL corresponding to link (JP,KP) has not been updated to show that it is no longer in the network. Hence, after loop "DØ 50" is ended, step 3 of the algorithm is completed by setting DP(JP,KP) = INFIN and DP(KP,JP) = INFIN. During loop "DØ 50," PP(JP,KP) was set to JP and PP(KP,JP) was set to KP since JP is in the M array and KP is in the N array.

Step 4 is accomplished in loop "DØ 80." In the coding, variable NØDES replaces n of the algorithm description. Otherwise, the coding follows step 4 in a straight forward manner. The program exits after printing debugging output, if this output was requested.

There is separate coding for the case when the link distance of (JP,KP) is 0. This coding begins at statement "200 MLIM = NØDES." Essentially, this coding performs steps 2, 3, and 4 more efficiently than the above coding. The reason is that in this case every node i, i = 1,2 ... NØDES is in both M and N. Step 2 is performed just by setting NLIM = NØDES, MLIM = NØDES and by loading every node number into arrays M and N. M and N must be defined because the subroutine calling LNKØUT may use them.

To perform step 3, instead of resetting elements of matrices DP and PP for each (i,j) in (M x N), reset DP and PP for all (i,j), i = 1 ... NØDES - 1, j = i+1 ... NØDES. This amounts to the same thing, but is more efficient. Most of step 3 is performed in loop "DØ 230." Check the use of MLEVEL and KFØRCE arrays as used in the coding of step 2 for link length of (JP,KP) > 0, as the same concept is used. Again, since MLEVEL had not been updated to show that link (JP,KP) is no longer constructed, after completing loop "DØ 230," the subroutine must set DP(JP,KP) = ∞ and DP(KP,J) = ∞, thereby completing step 3.

Step 4 is performed in loop "DØ 260." Instead of performing step 4, with each v = 1 ... NØDES, and for each $i \in M$ where $i \neq v$, and for each $\ell \in N$, the special nature of M and N makes it more efficient to do this for each v = 1,2 ... NØDES, for each i = 1, 2 ... NØDES where $i \neq v$, and for each ℓ = 1 ... NØDES. Note also that for a given v, all i for which DP(i,v) = ∞ may be excluded, for no updating of DP would occur anyway. The whole process could have been made more efficient for the case of link distances of (JP,KP) = 0 just by doing step 4 with v, v = 1 ... NØDES, and for each i, i = 1 ... NØDES - 1, where $i \neq v$ and DP(i,v) ≠ ∞, and for each ℓ = i + 1 ... NØDES.

Important arrays are:

- DP - an argument in the call to LNKØUT. Upon entry to LINKØUT, DP should be the INTEGER matrix of minimum internodal distances with link (JP,KP). Upon exit of LNKØUT, DP is the matrix of minimum internodal distances of the same network minus link (JP,KP). At all times the diagonal is zero.
- PP - an argument in the call to LNKØUT. Upon entry to LNKØUT, PP should be the INTEGER matrix of first back nodes for the network including link (JP,KP). Upon exit from LNKØUT, PP is the matrix of first back nodes of minimum paths for the network without link (JP,KP). At all times PP(I,I) = I, I = 1,2 ... NØDES.
- M - this array corresponds to set M in the algorithm described above.
- N - this array corresponds to set N in the above description.

The cross product (M x N) gives all node pairs whose minimum paths may have used link (JP,KP). A node pair {M(i), N(j)} in (M x N) would not also have {N(j), M(i)} in (M x N) unless link length (JP,KP) is zero. So, unless the link length (JP,KP) is 0, one must consider (M x N) and (N x M) to obtain all paths which could have changed.

MLEVEL - elements of this array tell whether a link is presently constructed or not, with one exception: the entry for link (JP,KP) still indicates it is constructed. See the description above for MLEVEL's use in this subroutine.

KFØRCE - a matrix indicating whether user has specified links to be forced or not. If KFØRCE(i,ℓ) < 0, link (i,ℓ) was specified by user not to be in the network at the budget constraint. If KFØRCE(i,ℓ) > 0, link (i,ℓ) was specified by user to be in the network at this budget constraint. If KFØRCE(i,ℓ) = 0, then link (i,ℓ) is not forced by user at this budget constraint. Check the description of the coding to see how KFØRCE is used in subroutine LNKØUT.

Important variables are:

JP - input to the subroutine; see description of KP.

KP - input to the subroutine; the link to be removed from the network is link (JP,KP). There is no restriction on the order of JP and KP.

JPRINT - input to the subroutine. This is a control for debugging output. The main algorithm often must remove a single link; however, it is important only at certain times to have the resultant minimum internodal distances and first back nodes printed. Variable JPRINT allows the program to print selectively at these important times. If JPRINT = 0, the debugging output will not be printed, even if IØ = 1. If JPRINT = 1, debugging output will be printed if IØ = 1, but will not be printed if IØ = 0.

SSTAR - link length of link (JP,KP), an INTEGER variable.

INFIN - integer used to represent infinity.

MLIM - number of elements in array M.

NLIM - number of elements in array N.

T - definition is clear in the coding; an INTEGER variable.

V - definition and use is clear from algorithm description and coding; an INTEGER variable.

For debugging output, much is printed. Assume that steps 1, 2, 3, and 4 must all be executed. Then, at the end of step 3, matrices DP and PP are printed in both cases, link length of (JP,KP) > 0 or link length of (JP,KP) = 0. First the message "LNKØUT" is printed, then matrix DP is printed at the end of step 3, and then matrix PP is printed at the end of step 3. Both DP and PP are printed one complete row at a time. DP is labelled "MATRIX ØF DISTANCES CØMPILED." PP is labelled "MATRIX ØF BACK NØDE CØMPILED."

At the completion of step 4, a summary of results is printed. The message "SUBRØUTINE LNKØUT" is printed, followed by the values of variables MLIM and NLIM and the link (JP,KP) which was removed from the network. Then the contents of array M and the contents of array N are printed. Matrix DP containing the new minimum distances is printed a row at a time with the label "MIN DISTANCE MATRIX." Matrix PP containing the new first back nodes is printed a complete row at a time with the label "1ST BACK NØDE MATRIX."

If at step 1, the original minimum distance between node JP and node KP was less than the link length, steps 2, 3 and 4 are not performed. As debugging output, the message "SUBRØUTINE LNKØUT" is printed followed by the printing of the link to be removed, its link length, and the original minimum distance between JP and KP. Then the message, "THEREFØRE REMØVING THE LINK CAUSES NØ CHANGES IN THE MIN PATH LENGTHS;" is printed. No other printing is done in this case since neither matrix DP nor matrix PP changed. The same output would occur if subroutine LNKØUT had been asked to remove link (JP,KP) with infinite link length.

LNKØUT prints some regular output regardless of the value of IØ or JPRINT, if an error occurs in the program. If LNKØUT is called with JP = KP, the program prints an error message upon detecting this condition, but will continue executing the main program. Nothing will have been done by LNKØUT, for this is checked for upon entry to the subroutine, the message is then printed, and the subroutine exits. Another error check is for (M ∩ N) not being the empty set, but the link length of (JP,KP) is greater than zero. An error message results because the algorithm will always have (M ∩ N) as the empty set when link length (JP,KP) = 0. It is not conceivable that such an error would occur. The main program would stop if this error occurs.

4.8.22 LINKIN(J,K,S,D,P,IPRINT)

This subroutine revises the minimum distance matrix and first back node matrix when a link is added to a network. For a definition of the problem and a statement of the algorithm implemented, see Section 4.2.1 . The algorithm is not restated here, but references to the steps as outlined in the above section are made. Almost all variables in the statement of the algorithm are the same as the names in the coding.

In the call to LINKIN (J,K,S,D,P, IPRINT); link (J,K) is the link to be added. Its length is S, an INTEGER variable; S takes the place of m of the statement of the algorithm. Upon entry, matrix D is the minimum distance matrix of the old network. Upon exit, matrix D has the minimum distances for the network plus link (J,K). Matrix P is the first back node matrix corresponding to the minimum distances of D. Both matrices D and P are the same as in the statement of the algorithm. They are both INTEGER arrays. Instead of n nodes as in the statement of the algorithm, there are NØDES nodes.

Step 1 is performed in the first 6 executable statements of the program. Note that the subroutine exits without making any changes in D or P if the link to be added has infinite length. Step 2 is performed in loop "DØ 10." Step 3 is performed in loop "DØ 20." Arrays M and N correspond to sets M and N of the statement of the algorithm. If either M or N is empty, step 4 is not performed. Step 4 is loop "DØ 40."

Important arrays are:

D - INTEGER matrix defined above.

P - INTEGER matrix defined above.

M - array representing set M in the statement of the algorithm. Node i ϵ M means that minimum path from i to k was shortened by using link (J,K).

N - array representing set N in the statement of the algorithm. Node ℓ ϵ N means the minimum path from j to ℓ was shortened by using link (J,K).

Important variables are:

IPRINT - binary input in call to LINKIN

0, do not print debugging output, even if IØ = 1;

1, print debugging output only if IØ = 1.

S - INTEGER variable defined above.

T - INTEGER variable used for temporary storage.

MI - MI = 1 at the end of step 3 means M is the empty set, so do not perform step 4.

NI - NI = 1 at the end of step 3 means N is the empty set, so do not perform step 4.

MLIM - MLIM = -1 means that no changes were made at all in the minimum paths because the link length was so large as to render it useless for the network constructed. If S = INFIN or S greater than or equal to the original minimum distance from J to K, MLIM = -1. Otherwise, MLIM is the number of nodes in array M.

NLIM - NLIM = -1 occurs if and only if MLIM = -1; see definition of MLIM. Otherwise, NLIM is the number of nodes in array N.

Debugging output is printed just before subroutine LINKIN exits. If link length of (J,K) is infinite, or if the link length of (J,K) is greater than or equal to the original minimum path length, the subroutine makes no computations. The debugging output consists of the message "SUBRØUTINE LINKIN," the link (J,K), its link length, the original minimum path length, and the message, "THEREFØRE NØ CHANGES ARE MADE IN THE MIN PATH LENGTHS."

If the link does actually change the minimum paths, the message "SUBRØUTINE LINKIN" is printed, followed by the values of MLIM and NLIM, and link (J,K). If M or N are nonempty arrays, their contents are printed. Matrix D is printed one complete row at a time, and matrix P is printed one row at a time. All debugging output occurs just before exit of the subroutine.

4.8.23 STAJ6B

This subroutine performs stages 6B, 6C, and 6D of the flowchart. The duties of STAJ6B are so similar to the duties of STAGE4, that the coding is almost identical. Reference to description of subroutine STAGE4 would be helpful.

The chief difference is that the link sought must allow enough budget within the budget constraint after its construction in order to construct at least one other link with MLEVEL = -1. Since subroutine STAGE6 checks exactly that, STAJ6B makes use of STAGE6. The values of IFØUN1, IFØUN2, and IFØUN3 must be saved in order to use STAGE6. Variable MX2 is initialized to (-1).

Loops 40 and 50 are basically identical to those of subroutine STAGE4 with the following exceptions:

1. IFØUN1, IFØUN2, IFØUN3 are temporarily defined to examine the candidate link by STAGE6 according to condition three of stage 6B in the flowchart;

2. MX2 rather than MX is used as the variable for maximizing the decrease in the sum of minimum internodal distances;

3. IFØUN4, IFØUN5, and IFØUN6 are used to find the best link, if there is one. If MX2 > -1, there is such a link, and the answer to stage 6C is YES. Therefore, MX is set to MX2, the decrease in sum of minimum distances for the best link of stage 6B. However, if MX2 = -1, there were no links satisfying stage 6B. In this case, since IFØUN4, IFØUN5, and IFØUN6 are never updated, with MX, they still hold the values determined at stage 4.

Important variables are:

IFØUN4 - see text above.

IFØUN5 - see text above.

IFØUN6 - see text above.

MX2 - see text above.

IF8 - a return variable,

0, go to stage 7 on exit;

1, go to stage 8 on exit.

NY - binary indicator,

0, the link can be constructed along with another link with MLEVEL = -1.

1, the link cannot be constructed with any other link with MLEVEL = -1.

All other variables and arrays are as defined earlier.

The debugging output is rather confusing. First a message is printed at the beginning to warn that the messages of subroutine STAGE6 are printed during subroutine STAJ6B. Therefore, STAGE6 messages indicate whether individual links passing conditions 1 and 2 of stage 6B also pass condition 3. The final message of 6B is the summary. If there is no link satisfying 6B, then the link selected by stage 4 is printed. The messages of STAJ6B are preceded and followed by a series of marks - /*/*/*/*/ ... - to separate this output from STAGE6 output.

BIBLIOGRAPHY

Anderson, T.W. (1968) Introduction to Multivariate Statistical Analysis, John Wiley and Sons, New York.

Beale, E.M.L. (1970) "Selecting an Optimal Subset," 451-562, in Integer and Nonlinear Programming, J. Abadie (ed.), North-Holland Publishing Company, Amsterdam.

Beale, E.M.L., M.G. Kendall and D.W. Mann (1967) "The Discarding of Variables in Multivariate Analysis," Biometrika, 54(3), 357-366.

Berge, C. and A. Ghouila-Houri (1965) Programming, Games and Transportation Networks, John Wiley and Sons, New York.

Boyce, D.E., A. Farhi and R. Weischedel (1969) "A Computer Program for Optimal Regression Analysis," RSRI Discussion Paper Series: No. 28, Regional Science Research Institute, Philadelphia.

———— (1970) "Using the Optimal Regression Program," Regional Science Department, University of Pennsylvania, Philadelphia.

———— (1973) "Optimal Network Problem: A Branch-and-Bound Algorithm," Environment and Planning, 5(4), 519-533.

Cooley, W.W. and P.R. Lohnes (1962) Multivariate Procedures for the Behavioral Sciences, John Wiley and Sons, New York.

Dantzig, G.B. (1966) "All Shortest Routes in a Graph," Technical Report 66-3, Operations Research House, Stanford University, California.

Dixon, W.J. (1968) BMD Biomedical Computer Programs, University of California Press, Berkeley and Los Angeles, California.

Draper, N.R. and H. Smith (1966) Applied Regression Analysis, John Wiley and Sons, New York.

Dreyfus, S.E. (1969) "An Appraisal of Some Shortest-Path Algorithms," Operations Research, 17, 395-412.

Efroymson, M.A. (1962) "Multiple Regression Analysis," in A. Ralston and H.S. Wilf (eds.) Mathematical Methods for Digital Computers, John Wiley and Sons, New York.

Floyd, R.W. (1962) "Algorithm 97, Shortest Path," Communications of the ACM, 5, 345.

Halder, A.K. (1970) "The Method of Competing Links," Transportation Science, 4, 37-51.

Hickey, R.J., D.E. Boyce, E.B. Harner and R.C. Clelland (1970a) "Ecological Statistical Studies Concerning Environmental Pollution and Chronic Disease," IEEE Transactions on Geoscience Electronics, GE-8, 186-202; reproduced with minor editing in Chapter 11, "Environmental Pollution and Human Health," B.J.L. Berry and F.E. Horton (eds.) Urban Environment Management: Planning for Pollution Control, Prentice-Hall, Englewood Cliffs, N.J., 1974.

——— (1970b) "Ecological Statistical Studies on Environmental Pollution and Chronic Disease in Metropolitan Areas of the United States," Discussion Paper Series No. 35, Regional Science Research Institute, Philadelphia.

——— (1971) Exploratory Ecological Studies of Variables Related to Chronic Disease Mortality Rates, University of Pennsylvania, Philadelphia.

Hoang, H.H. (1973) "A Computational Approach to the Selection of an Optimal Network," Management Science, 19(5), 488-498.

Morrison, D.F. (1967) Multivariate Statistical Methods, McGraw-Hill Book Company, New York.

Murchland, J.D. (1965) "A New Method for Finding All Elementary Paths in a Complete Directed Graph," LSE-TNT-22, London School of Economics, London.

——— (1967) "The Effect of Increasing or Decreasing the Length of a Single Arc on All Shortest Path Distances in a Graph," LBS-TNT-26, London Business School, London.

——— (1970) A Fixed Method for All Shortest Distances in a Directed Graph and for the Inverse Problem, Ph.D. Dissertation, University of Karlsruhe, Karlsruhe, Germany.

Olkin, I., and J.W. Pratt (1958) "Unbiased Estimation of Certain Correlation Coefficients," Annals of Mathematical Statistics, 29, 201-211.

Sonquist, J.A., and J.N. Morgan (1970) The Detection of Interaction Effects, Institute for Social Research, University of Michigan, Ann Arbor, Michigan.

Steenbrink, P.A. (1974a) Optimization of Transport Networks, John Wiley and Sons, New York.

——— (1974b) "Transport Network Optimization in the Dutch Integral Transportation Study," Transportation Research, 8, 11-27.

Index

Note: computer programs are indicated as follows:
R - optimal regression analysis
I - interdependence analysis
N - optimal network analysis

Vol. 59: J. A. Hanson, Growth in Open Economics. IV, 127 pages. 4°. 1971. DM 16,–

Vol. 60: H. Hauptmann, Schätz- und Kontrolltheorie in stetigen dynamischen Wirtschaftsmodellen. V, 104 Seiten. 4°. 1971. DM 16,–

Vol. 61: K. H. F. Meyer, Wartesysteme mit variabler Bearbeitungsrate. VII, 314 Seiten. 4°. 1971. DM 24,–

Vol. 62: W. Krelle u. G. Gabisch unter Mitarbeit von J. Burgermeister, Wachstumstheorie. VII, 223 Seiten. 4°. 1972. DM 20,–

Vol. 63: J. Kohlas, Monte Carlo Simulation im Operations Research. VI, 162 Seiten. 4°. 1972. DM 16,–

Vol. 64: P. Gessner u. K. Spremann, Optimierung in Funktionenräumen. IV, 120 Seiten. 4°. 1972. DM 16,–

Vol. 65: W. Everling, Exercises in Computer Systems Analysis. VIII, 184 pages. 4°. 1972. DM 18,–

Vol. 66: F. Bauer, P. Garabedian and D. Korn, Supercritical Wing Sections. V, 211 pages. 4°. 1972. DM 20,–

Vol. 67: I. V. Girsanov, Lectures on Mathematical Theory of Extremum Problems. V, 136 pages. 4°. 1972. DM 16,–

Vol. 68: J. Loeckx, Computability and Decidability. An Introduction for Students of Computer Science. VI, 76 pages. 4°. 1972. DM 16,–

Vol. 69: S. Ashour, Sequencing Theory. V, 133 pages. 4°. 1972. DM 16,–

Vol. 70: J. P. Brown, The Economic Effects of Floods. Investigations of a Stochastic Model of Rational Investment Behavior in the Face of Floods. V, 87 pages. 4°. 1972. DM 16,–

Vol. 71: R. Henn und O. Opitz, Konsum- und Produktionstheorie II. V, 134 Seiten. 4°. 1972. DM 16,–

Vol. 72: T. P. Bagchi and J. G. C. Templeton, Numerical Methods in Markov Chains and Bulk Queues. XI, 89 pages. 4°. 1972. DM 16,–

Vol. 73: H. Kiendl, Suboptimale Regler mit abschnittweise linearer Struktur. VI, 146 Seiten. 4°. 1972. DM 16,–

Vol. 74: F. Pokropp, Aggregation von Produktionsfunktionen. VI, 107 Seiten. 4°. 1972. DM 16,–

Vol. 75: GI-Gesellschaft für Informatik e.V. Bericht Nr. 3. 1. Fachtagung über Programmiersprachen · München, 9–11. März 1971. Herausgegeben im Auftag der Gesellschaft für Informatik von H. Langmaack und M. Paul. VII, 280 Seiten. 4°. 1972. DM 24,–

Vol. 76: G. Fandel, Optimale Entscheidung bei mehrfacher Zielsetzung. 121 Seiten. 4°. 1972. DM 16,–

Vol. 77: A. Auslender, Problemes de Minimax via l'Analyse Convexe et les Inégalités Variationnelles: Théorie et Algorithmes. VII, 132 pages. 4°. 1972. DM 16,–

Vol. 78: GI-Gesellschaft für Informatik e.V. 2. Jahrestagung, Karlsruhe, 2.–4. Oktober 1972. Herausgegeben im Auftrag der Gesellschaft für Informatik von P. Deussen. XI, 576 Seiten. 4°. 1973. DM 36,–

Vol. 79: A. Berman, Cones, Matrices and Mathematical Programming. V, 96 pages. 4°. 1973. DM 16,–

Vol. 80: International Seminar on Trends in Mathematical Modelling, Venice, 13–18 December 1971. Edited by N. Hawkes. VI, 288 pages. 4°. 1973. DM 24,–

Vol. 81: Advanced Course on Software Engineering. Edited by F. L. Bauer. XII, 545 pages. 4°. 1973. DM 32,–

Vol. 82: R. Saeks, Resolution Space, Operators and Systems. X, 267 pages. 4°. 1973. DM 22,–

Vol. 83: NTG/GI-Gesellschaft für Informatik, Nachrichtentechnische Gesellschaft. Fachtagung „Cognitive Verfahren und Systeme", Hamburg, 11.–13. April 1973. Herausgegeben im Auftrag der NTG/GI von Th. Einsele, W. Giloi und H.-H. Nagel. VIII, 373 Seiten. 4°. 1973. DM 28,–

Vol. 84: A. V. Balakrishnan, Stochastic Differential Systems I. Filtering and Control. A Function Space Approach. V, 252 pages. 4°. 1973. DM 22,–

Vol. 85: T. Page, Economics of Involuntary Transfers: A Unified Approach to Pollution and Congestion Externalities. XI, 159 pages. 4°. 1973. DM 18,–

Vol. 86: Symposium on the Theory of Scheduling and Its Applications. Edited by S. E. Elmaghraby. VIII, 437 pages. 4°. 1973. DM 32,–

Vol. 87: G. F. Newell, Approximate Stochastic Behavior of n-Server Service Systems with Large n. VIII, 118 pages. 4°. 1973. DM 16,–

Vol. 88: H. Steckhan, Güterströme in Netzen. VII, 134 Seiten. 4°. 1973. DM 16,–

Vol. 89: J. P. Wallace and A. Sherret, Estimation of Product Attributes and Their Importances. V, 94 pages. 4°. 1973. DM 16,–

Vol. 90: J.-F. Richard, Posterior and Predictive Densities for Simultaneous Equation Models. VI, 226 pages. 4°. 1973. DM 20,–

Vol. 91: Th. Marschak and R. Selten, General Equilibrium with Price-Making Firms. XI, 246 pages. 4°. 1974. DM 22,–

Vol. 92: E. Dierker, Topological Methods in Walrasian Economics. IV, 130 pages. 4°. 1974. DM 16,–

Vol. 93: 4th IFAC/IFIP International Conference on Digital Computer Applications to Process Control, Zürich/Switzerland, March 19–22, 1974. Edited by M. Mansour and W. Schaufelberger. XVIII, 544 pages. 4°. 1974. DM 36,–

Vol. 94: 4th IFAC/IFIP International Conference on Digital Computer Applications to Process Control, Zürich/Switzerland, March 19–22, 1974. Edited by M. Mansour and W. Schaufelberger. XVIII, 546 pages. 4°. 1974. DM 36,–

Vol. 95: M. Zeleny, Linear Multiobjective Programming. XII, 220 pages. 4°. 1974. DM 20,–

Vol. 96: O. Moeschlin, Zur Theorie von Neumannscher Wachstumsmodelle. XI, 115 Seiten. 4°. 1974. DM 16,–

Vol. 97: G. Schmidt, Über die Stabilität des einfachen Bedienungskanals. VII, 147 Seiten. 4°. 1974. DM 16,–

Vol. 98: Mathematical Methods in Queueing Theory. Proceedings of a Conference at Western Michigan University, May 10–12, 1973. Edited by A. B. Clarke. VII, 374 pages. 4°. 1974. DM 28,–

Vol. 99: Production Theory. Edited by W. Eichhorn, R. Henn, O. Opitz, and R. W. Shephard. VIII, 386 pages. 4°. 1974. DM 32,–

Vol. 100: B. S. Duran and P. L. Odell, Cluster Analysis. A Survey. VI, 137. 4. 1974. DM 18,–

Vol. 101: W. M. Wonham, Linear Multivariable Control. A Geometric Approach. X, 344 Seiten. 4°. 1974. DM 30,–

Vol. 103: D. E. Boyce, A. Farhi, and R. Weischedel, Optimal Subset Selection. Multible Regression, Interdependence and Optimal Network Algorithms. XIII, 187 pages. 4°. 1974. DM 20,–